T0073088

SpringerBriefs in Computer Science

SpringerBriefs present concise summaries of cutting-edge research and practical applications across a wide spectrum of fields. Featuring compact volumes of 50 to 125 pages, the series covers a range of content from professional to academic.

Typical topics might include:

- A timely report of state-of-the art analytical techniques
- A bridge between new research results, as published in journal articles, and a contextual literature review
- A snapshot of a hot or emerging topic
- An in-depth case study or clinical example
- A presentation of core concepts that students must understand in order to make independent contributions

Briefs allow authors to present their ideas and readers to absorb them with minimal time investment. Briefs will be published as part of Springer's eBook collection, with millions of users worldwide. In addition, Briefs will be available for individual print and electronic purchase. Briefs are characterized by fast, global electronic dissemination, standard publishing contracts, easy-to-use manuscript preparation and formatting guidelines, and expedited production schedules. We aim for publication 8–12 weeks after acceptance. Both solicited and unsolicited manuscripts are considered for publication in this series.

**Indexing: This series is indexed in Scopus, Ei-Compendex, and zbMATH **

Acknowledgments

The authors of this book would like to acknowledge our editors for their help and guidance. All the supports, advice, and encouragement from peer experts, scholars, students, and team members at the Beijing Institute of Technology are remarkably meaningful and invaluable for the accomplishment of this book. The authors sincerely appreciate all the individuals and organizations who reinforce and raise the quality of this book. This book is partially supported by the National Natural Science Foundation of China under Grant No. 62102028, 62072040, 62172038, and 62202051.

Contents

Chapter 1
A Brief Introduction

Abstract In this chapter, we first introduce the background regarding Mobile Crowdsensing (MCS) and present an overview of MCS. Then, we specifically state the incentive mechanism design problem for MCS. Finally, we demonstrate the book structure for convenience.

Keywords Mobile crowdsensing · Architecture · Incentive mechanism · Game theory

1.1 Overview of Mobile Crowdsensing

Recently, the proliferation and prevalence of mobile devices enable a new kind of sensing paradigm, mobile crowdsensing (MCS) [1, 2], which allows a platform to collect data from on-site users carrying mobile devices anywhere and anytime. Therefore, MCS has revolutionized the traditional sensing paradigm (e.g., Wireless Sensor Networks, WSN) in the Internet of Things (IoT), which enables a large number of successful crowdsensing applications that cover and affect people's daily lives.

Typically, the architecture of MCS is illustrated in Fig. 1.1 where an MCS system is made up of several requesters, a platform and a large number of mobile users. These three parties interact in the MCS system:

- Requester: An entity that has sensing demand and generates sensing tasks over time. Note that we focus on a scenario where multiple requesters generate tasks;
- User: A task actuator who has sensing and computing capability to perform tasks. There are multiple users working in the MCS system;[1]
- Platform: An intermediate who collaborates the interaction between requesters and users (i.e., receives tasks from requesters and allocates them to users).

---

[1] As shown in Fig. 1.1, the user can be a vehicle with general-purpose sensors.

Y. Li et al., *Incentive Mechanism for Mobile Crowdsensing*, SpringerBriefs in Computer Science, https://doi.org/10.1007/978-981-99-6921-0_1

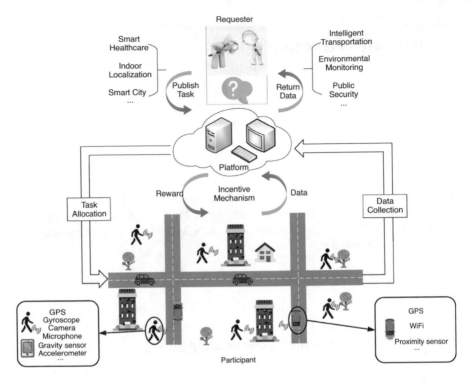

Fig. 1.1 Illustration of crowdsensing architecture

Generally speaking, downstream applications like healthcare [3–5], smart transportation [6–8], smart city [9–11], environmental monitoring [12, 13], accurate localization [14–16], social networking [17] are developed based on the data-related services. These applications are usually time-sensitive and location-aware. To guarantee applications' quality of service (QoS), real-time and location-aware sensing data should be collected. However, on the one hand, the service provider (i.e., requester) cannot perform the data collection individually. On the other hand, current mobile devices (e.g. smartphones, tablets, wearable equipment) grow ever-increasingly. Moreover, mobile devices are commonly held by people anywhere and anytime and are increasingly embedded with various sensors (accelerometer, thermometer, gyroscope, GPS, microphone, and camera). If all kinds of "super sensors" together contribute to crowdsensing, it will be the unprecedented sensing network in the world. Realizing the potential of crowdsensing, a platform is giving rise to serve the requesters by harnessing the power of the crowd to collectively complete large-scale sensing tasks. Therefore, MCS streamlines data collection and enhances the value of easily-generated data from mobile users.

From Fig. 1.1, we can observe that task allocation (or user selection), result aggregation (or quality management) and incentive mechanism are three main studied problems in this area. Among them, designing an effective incentive mechanism is of paramount importance to the success of MCS because incentive mechanisms ensure users' participation such that task allocation (or user selection) could happen. In fact, task allocation (or user selection) is often a sub-problem of incentive mechanisms, and result aggregation (or quality management) can be considered in incentive mechanisms. Therefore, this book focuses on designing incentive mechanisms for the MCS.

1.2   Incentive Mechanism for Mobile Crowdsensing

The feasibility and success of crowdsensing highly rely on the number of users who participate in performing sensing tasks, as well as the quality of the collected sensing data. The early works assume that users voluntarily participate in performing tasks [18–20]. However, performing sensing tasks incurs some costs for users. On the one hand, performing sensing tasks requires the built-in sensors to sense the related data, which consumes devices' energy resources (e.g., battery) and computing resources. Returning the results of sensing tasks leads to the potential communication cost to users. On the other hand, sensing tasks are usually related to specific locations. Performing tasks will reveal users' location information, which causes privacy concerns to users. Therefore, users are unwilling to participate in tasks unless they are sufficiently rewarded to compensate the incurred cost. But excessively offering rewards will impose more budget for the platform, thus it is necessary to address how to design rational and appropriate rewards to elicit users' contributions for sensing tasks. This challenging interaction between the platform and users involves an incentive mechanism design problem.

The incentive mechanisms for crowdsensing are divided into non-monetary incentive mechanisms and monetary incentive mechanisms. Non-monetary incentive mechanisms incentivize users to participate in crowdsensing for performing tasks by intrinsic rewards like gamification [21–23], virtual credit [24, 25] and self-interest [4]. Monetary incentive mechanisms directly use the money to compensate users' sensing costs. Therefore, the monetary incentive mechanism design problem is related to addressing a pricing problem that determines how much money is appropriate to incentivize users. In this book, we pay attention to investigating monetary incentive mechanisms since the pricing problem can be well modeled by game theory [26, 27].

Game theory is leveraged to address the situation where the payoff of one agent is not only dependent on its strategy but also on the others' strategies. It analyzes and predicts the outcome of the game by assuming that agents are rational and selfish to interact with each other in a game. In crowdsensing, the platform and users have conflicting interests. The platform expects to offer as fewer rewards to solicit more participation as possible for higher revenues. However, the users anticipate more rewards for maximizing their utility. The utility of the platform relies on how many sensing tasks participants complete and how many rewards the platform offers. The utility of each user is determined by how much rewards the user receives and how much cost the user incurs from performing tasks. The rewards the user receives are even affected by other users' decisions due to their competition to perform tasks. Game theory provides an insightful solution to accommodate the conflicting interaction between the platform and users.

Designing an efficient and effective incentive mechanism poses some challenges: (1) limited resources and budget. Sensing tasks require the usage of sensors on users' mobile devices to collect data, which consumes battery energy and data plan to transmit the collected data. Besides, users cannot aggregate and process the collected data with limited computing resources. For the platform, it has a constrained budget to offer rewards due to limited revenues [28]. (2) rationality and selfishness of decision-makers. Both the platform and users only concentrate on unilaterally maximizing their own payoffs. Their strategic actions influence the quality of sensing service and degenerate the performance of crowdsensing. For example, after receiving rewards, users do not perform tasks or collect low-quality data [29–31]. (3) asymmetry of the information about the crowdsensing system (to be described in detail shortly). The information (e.g., cost, valuation, distribution, etc.) of each user is private, which makes users' participating decisions and interaction between users and platform more complicated [29, 32–37].

The above challenges essentially involve the pricing problem, which is about rewards the platform needs to set for each task [38–40] or each user [30, 31, 41] to incentivize participation. To address the pricing problem, most existing works turn to game theory due to its convenience and suitability of modeling complex interactions among multi-agents.

While many existing papers study incentive mechanisms in MCS, to the best of our knowledge, there are few books giving a comprehensive review of the incentive mechanism for MCS, especially from the game-theoretic perspective. This book aims to fill this void. In this book, we introduce our works on designing incentive mechanisms for MCS from different perspectives under different settings, which produces: (1) a long-term incentive mechanism for MCS, (2) a fair incentive mechanism for MCS, (3) a collaborative incentive mechanism for MCS, (4) a coopetition-aware incentive mechanism for MCS.

1.3 Book Structure

This book mainly investigates the incentive mechanism design problem from the game-theoretic perspective. The remainder of this book is organized as follows:

Chapter 2 investigates the long-term incentive mechanism design problem for MCS under a time-varying scenario where the platform aims to continuously incentivize users and ensure requesters' profits in order to guarantee the sustainment of MCS systems in the long term. We utilize a three-stage Stackelberg game to modeling the three-stage interactions between the platform and users in each round. Moreover, we incorporate Lyapunov optimization into the Stackelberg game to address online decision problems for the platform such that long-term selection constraints and time-average profits can be ensured.

Chapter 3 investigates the fair incentive mechanism design problem for MCS under a time-varying scenario where users' sensing qualities are unknown a priori. In this setting, we aim to address the problems in the strategic interactions: (1) How to recruit a high-quality participant when the quality of information is unknown; (2) How to ensure fairness in the user selection stage; (3) How to preserve privacy for the selected users due to the underlying privacy leakage during data aggregation. In this chapter, we jointly address these issues and propose the three-stage Stackelberg-based incentive mechanism for the platform to recruit participants. In detail, we leverage combinatorial volatile multi-armed bandits (CVMAB) to elicit unknown users' sensing qualities. We use the drift-plus-penalty (DPP) technique in Lyapunov optimization to handle the fairness requirements. We blur the quality feedback with tunable Laplacian noise such that the incentive mechanism protects locally differential privacy (LDP).

Chapter 4 investigates the collaborative incentive mechanism design problem for MCS under cooperation with an additional POI-tagging App (e.g., Pokemon Go). In this setting, the strategic interactions become three-stage and complicate the previous two-party incentive mechanisms' interactions. We model the three-stage interactions between users, platform, and App by leveraging the three-stage Stackelberg game. App first determines the POI-tagging price to maximize its payoff. Platform and users subsequently decide how to determine tasks reward and select edges to be tagged, and how to select the best task to perform, respectively. Under the collaboration of the POI-tagging App, the platform can ensure the participation rate for each task such that the sensing services can be continuously provided to the requester.

Chapter 5 investigates the coopetition-aware incentive mechanism design problem for MCS under multiple platforms scenario. In the multiple platforms setting, there are two interactions between different platforms: competitive platforms and cooperative platforms, which are also regarded as coopetition-aware interactions. In the competitive platform scenario, platforms decide their prices on rewards to attract more participants, while the users choose which platform to work for. We model such a competitive platform scenario as a two-stage Stackelberg game. In the cooperative platform scenario, platforms cooperate to share sensing data with each other. We model it as many-to-many bargaining. By analyzing the Stackelberg Equilibrium (SE) and Nash Bargaining Solution (NBS), we obtain the design principle of incentive mechanisms both for the competitive platform scenario and the cooperative platform scenario.

Chapter 6 summarizes the book and discusses future directions in this area.

References

1. Lane, N.D., Miluzzo, E., Lu, H., Peebles, D., Choudhury, T., Campbell, A.T.: A survey of mobile phone sensing. IEEE Commun. Mag. **48**(9) (2010)
2. Ganti, R.K., Ye, F., Lei, H.: Mobile crowdsensing: current state and future challenges. IEEE Commun. Mag. **49**(11) (2011)
3. Consolvo, S., McDonald, D.W., Toscos, T., Chen, M.Y., Froehlich, J., Harrison, B., Klasnja, P., LaMarca, A., LeGrand, L., Libby, R., et al.: Activity sensing in the wild: a field trial of ubifit garden. In: Proceedings of the SIGCHI Conference on Human Factors in Computing Systems, pp. 1797–1806. ACM (2008)
4. Reddy, S., Parker, A., Hyman, J., Burke, J., Estrin, D., Hansen, M.: Image browsing, processing, and clustering for participatory sensing: lessons from a dietsense prototype. In: Proceedings of the 4th Workshop on Embedded Networked Sensors, pp. 13–17. ACM (2007)
5. Zhang, X., Li, W., Chen, X., Lu, S.: Moodexplorer: Towards compound emotion detection via smartphone sensing. Proc. ACM Interact. Mob. Wearable Ubiquitous Technol. **1**(4), 176 (2018)
6. Thiagarajan, A., Ravindranath, L., LaCurts, K., Madden, S., Balakrishnan, H., Toledo, S., Eriksson, J.: Vtrack: accurate, energy-aware road traffic delay estimation using mobile phones. In: Proceedings of the 7th ACM Conference on Embedded Networked Sensor Systems, pp. 85–98. ACM (2009)
7. Hunter, T., Teodor, M., Matei, Z., Samy, M., Justin, M., Michael, J.F., Pieter, A., Alexandre, M.B.: Scaling the mobile millennium system in the cloud. In: Proceedings of the 2nd ACM Symposium on Cloud Computing, pp. 1–8. (2011)
8. Nawaz, S., Efstratiou, C., Mascolo, C.: Parksense: A smartphone based sensing system for on-street parking. In: ACM MobiCom (2013)
9. Kim, S., Robson, C., Zimmerman, T., Pierce, J., Haber, E.M.: Creek watch: pairing usefulness and usability for successful citizen science. In: ACM CHI (2011)
10. Yabe, T., Tsubouchi, K., Sekimoto, Y.: Cityflowfragility: measuring the fragility of people flow in cities to disasters using gps data collected from smartphones. Proc. ACM Interact. Mob. Wearable Ubiquitous Technol. **1**(3), 117 (2017)
11. Lendák, I.: Mobile crowd-sensing in the smart city. European Handbook of Crowdsourced Geographic Information, vol. 353. Ubiquity Press (2015)
12. Mun, M., Reddy, S., Shilton, K., Yau, N., Burke, J., Estrin, D., Hansen, M., Howard, E., West, R., Boda, P.: Peir, the personal environmental impact report, as a platform for participatory sensing systems research. In: Proceedings of the 7th International Conference on Mobile Systems, Applications, and Services, pp. 55–68. ACM (2009)
13. Zheng, Y., Liu, F., Hsieh, H.-P.: U-air: when urban air quality inference meets big data. In: ACM SIGKDD, pp. 1436–1444 (2013)
14. Rai, A., Chintalapudi, K.K., Padmanabhan, V.N., Sen, R.: Zee: Zero-effort crowdsourcing for indoor localization. In: Proceedings of the 18th Annual International Conference on Mobile Computing and Networking, pp. 293–304. ACM (2012)
15. Chen, H., Li, F., Hei, X., Wang, Y.: Crowdx: Enhancing automatic construction of indoor floorplan with opportunistic encounters. Proc. ACM Interact. Mob. Wearable Ubiquitous Technol. **2**(4), 159 (2018)
16. Gao, R., Zhao, M., Ye, T., Ye, F., Wang, Y., Bian, K., Wang, T., Li, X.: Jigsaw: Indoor floor plan reconstruction via mobile crowdsensing. In: ACM MobiCom (2014)
17. Li, J., Cai, Z., Yan, M., Li, Y.: Using crowdsourced data in location-based social networks to explore influence maximization. In: IEEE International Conference on Computer Communications (INFOCOM), pp. 1–9 (2016)
18. Hull, B., Bychkovsky, V., Zhang, Y., Chen, K., Goraczko, M., Miu, A., Shih, E., Balakrishnan, H., Madden, S.: Cartel: a distributed mobile sensor computing system. In: Proceedings of the 4th International Conference on Embedded Networked Sensor Systems, pp. 125–138. ACM (2006)

19. Mathur, S., Jin, T., Kasturirangan, N., Chandrasekaran, J., Xue, W., Gruteser, M., Trappe, W.: Parknet: drive-by sensing of road-side parking statistics. In: Proceedings of the 8th International Conference on Mobile Systems, Applications, and Services, pp. 123–136. ACM (2010)

20. Zhao, D., Li, X.-Y., Ma, H.: How to crowdsource tasks truthfully without sacrificing utility: Online incentive mechanisms with budget constraint. In: IEEE Conference on Computer Communications (INFOCOM), pp. 1213–1221 (2014)

21. Barkhuus, L., Chalmers, M., Tennent, P., Hall, M., Bell, M., Sherwood, S., Brown, B.: Picking pockets on the lawn: the development of tactics and strategies in a mobile game. In: International Conference on Ubiquitous Computing, pp. 358–374. Springer (2005)

22. Schlieder, C., Kiefer, P., Matyas, S.: Geogames: Designing location-based games from classic board games. IEEE Intell. Syst. 21(5), 40–46 (2006)

23. Bell, M., Reeves, S., Brown, B., Sherwood, S., MacMillan, D., Ferguson, J., Chalmers, M.: Eyespy: supporting navigation through play. In: Proceedings of the SIGCHI Conference on Human Factors in Computing Systems, pp. 123–132. ACM (2009)

24. Lan, K.-C., Chou, C.-M., Wang, H.-Y.: An incentive-based framework for vehicle-based mobile sensing. Procedia Comput. Sci. 10, 1152–1157 (2012)

25. Chou, C.-M., Lan, K.-C., Yang, C.-F.: Using virtual credits to provide incentives for vehicle communication. In: 2012 12th International Conference on ITS Telecommunications, pp. 579–583. IEEE (2012)

26. Fudenberg, D., Tirole, J.: Game theory. Technical Report, The MIT Press, 1991

27. Nisan, N., Roughgarden, T., Tardos, E., Vazirani, V.V.: Algorithmic Game Theory. Cambridge University Press, Cambridge (2007)

28. Jaimes, L.G., Vergara-Laurens, I., Labrador, M.A.: A location-based incentive mechanism for participatory sensing systems with budget constraints. In: 2012 IEEE International Conference on Pervasive Computing and Communications (PerCom), pp. 103–108. IEEE (2012)

29. Gao, L., Hou, F., Huang, J.: Providing long-term participation incentive in participatory sensing. In: IEEE INFOCOM (2015)

30. Yang, D., Xue, G., Fang, X., Tang, J.: Crowdsourcing to smartphones: incentive mechanism design for mobile phone sensing. In: ACM MobiCom (2012)

31. Yang, D., Xue, G., Fang, X., Tang, J.: Incentive mechanisms for crowdsensing: Crowdsourcing with smartphones. IEEE/ACM Trans. Netw. (TON) 24(3), 1732–1744 (2016)

32. Peng, D., Wu, F., Chen, G.: Pay as how well you do: A quality based incentive mechanism for crowdsensing. In: Proceedings of the 16th ACM International Symposium on Mobile Ad Hoc Networking and Computing, pp. 177–186. ACM (2015)

33. Luo, T., Tan, H.-P., Xia, L.: Profit-maximizing incentive for participatory sensing. In: IEEE INFOCOM, pp. 1–9 (2014)

34. Yang, G., He, S., Shi, Z., Chen, J.: Promoting cooperation by the social incentive mechanism in mobile crowdsensing. IEEE Commun. Mag. 55(3), 86–92 (2017)

35. Luo, T., Kanhere, S.S., Das, S.K., TAN Hwee-Pink: Incentive mechanism design for heterogeneous crowdsourcing using all-pay contests. IEEE Trans. Mob. Comput. 15(9), 2234 (2016)

36. Li, M., Lin, J., Yang, D., Xue, G., Tang, J.: Quac: quality-aware contract-based incentive mechanisms for crowdsensing. In: 2017 IEEE 14th International Conference on Mobile Ad Hoc and Sensor Systems (MASS), pp. 72–80 (2017)

37. Zhao, N., Fan, M., Tian, C., Fan, P.: Contract-based incentive mechanism for mobile crowdsourcing networks. Algorithms 10(3), 104 (2017)

38. DiPalantino, D., Vojnovic, M.: Crowdsourcing and all-pay auctions. In: Proceedings of the 10th ACM conference on Electronic commerce, pp. 119–128 (2009)

39. He, S., Shin, D.-H., Zhang, J., Chen, J.: Toward optimal allocation of location dependent tasks in crowdsensing. In: IEEE INFOCOM, pp. 745–753 (2014)

40. Zhang, Y., Gu, Y., Pan, M., Tran, N.H., Dawy, Z., Han, Z.: Multi-dimensional incentive mechanism in mobile crowdsourcing with moral hazard. IEEE Trans. Mob. Comput. **17**(3), 604–616 (2018)
41. Lee, J.-S., Hoh, B.: Dynamic pricing incentive for participatory sensing. Pervasive Mob. Comput. **6**(6), 693–708 (2010)

Chapter 2
Long-Term Incentive Mechanism for Mobile Crowdsensing

Abstract In this chapter, we propose an incentive mechanism for crowdsensing under the continuous and time-varying scenario using a three-stage Stackelberg game. In such a scenario, different requesters generate sensing tasks with payments to the platform at each time slot. The platform makes pricing decisions to determine rewards for tasks without complete information, and then notifies task-price pairs to online users in Stage I. In Stage II, users select optimal tasks as their interests under certain constraints and report back to the platform. The platform fairly selects users as workers in order to ensure users' long-term participation in Stage III. We use Lyapunov optimization to address online decision problems for the platform in Stage I and III where there are no prior knowledge and future information available. We propose an FPTAS for users to derive their interests of tasks based on their mobile devices' computing capabilities in Stage II. Numerical results in simulations validate the significance and superiority of our proposed incentive mechanism.

Keywords Multiple-round crowdsensing · Long-term constraints · Three-stage Stackelberg game

2.1 Introduction

2.1.1 Motivations

In recent years, the emerging mobile era brings pervasive and prevalent smart devices (e.g., mobile phones, tablets and wearable watches, etc.) into modern society, which has revolutionized human's life in every aspect. Especially making full use of sensors built-in users' mobile phone to conduct real-time measurements about the physical world has given rise to a new sensing paradigm in IoT area, called *Crowdsensing* [1, 2]. The successful cases of crowdsensing include Waze [3] for real-time traffic monitoring, Gigwalk [4] for mobile market research, U-Air [5] and CCS-TA [6] for air quality monitoring, and FlierMeet [7] for public information sharing.

To build a good ecosystem for crowdsensing, a well-designed incentive mechanism is necessary and indispensable [8]. The incentive mechanism must satisfy the
following three key requirements to promote sustainability and achieve the success
of crowdsensing: (1) the high-quality completeness of requesters' tasks; (2) users'
extensive participation in tasks; (3) profitability for the platform. Unfortunately,
some existing works [9–11] relax some of the above requirements and impose
some strong assumptions. On the one hand, they focus on either whether tasks
are allocated to users for performing [9] or how much sensing and computing
resources users contribute to perform tasks [10, 11]. They both neglect the quality
issue of tasks to be completed. On the other hand, these works suppose only
one requester posts task(s) to the platform and their schemes only consider static
situations. Actually, different tasks generate over time. Thus, interactions over
multiple time slots should be taken into account. Note that, these existing works
cannot be easily extended to the multiple time-slots scenario due to the dynamics
of requesters and users (i.e., various tasks demand, mobility and "sleep" of users).
Although a later work [12] proposes a quality-driven incentive mechanism, it fails
to handle the dynamics of requesters and users. To this end, we pay attention
to a practical time-varying dynamic scenario where (1) multiple time slots are
considered; (2) requesters generate tasks over time; and (3) users may sleep due to
low battery and do not join crowdsensing in some slots. This kind of scenario is more
practical in many applications. For example, a set of requesters submit task demands
with various payments to a crowdsourcing platform like Amazon Mechanical Turk,
while the platform recruits a set of active mobile users to perform the data collection
tasks for requesters at the current time slot. In the research community, a similar
scenario is increasingly attracting more and more attention. For instance, Wei et al.
in [13] propose a double auction-based incentive mechanism to model the dynamic
interaction between service users and service providers at different time slots. Jin
et al. [14] propose an incentive mechanism called CENTURION to model the
interaction between multi-requester and multi-user using a double auction. As a toy
example, a service user wants to collect some photos at a set of specific points of
interest for training a deep learning model. Then, the service user can submit his/her
request to the platform that can recruit a set of active users as service providers to
take photos at the required points of interest. Different task demands of different
service users vary over different time slots and different users are active at different
time slots.

2.1.2 Challenges

In such a scenario, we aim to guarantee the fairness of users to be selected (i.e.,
scheduled) as workers and the platform's time-average profit, such that users will
stay at the crowdsensing system and continuously participate in performing tasks,
and hence sustainability of crowdsensing is achieved due to the guaranteeing

profitability for the platform. However, designing such an incentive mechanism to achieve the above two goals exists several unique challenges as discussed below.

Uncertainty of Tasks Each requester has various task demands over time. This implies that different requesters generate tasks to the platform at each time slot. Additionally, the requesters who generate tasks have different budgets and offer different payments for their task subscriptions. Thus, the uncertainty of tasks adds difficulty to the static incentive mechanisms in the existing literature.

Dynamics of Users Each user has evolving statuses over time. Their locations are changing according to users' mobility model and the remaining batteries are consumed due to the activities on their mobile phones. As a result, users' sensing costs (i.e., the cost incurred by the distance between the user and the task) are different at each time slot. Moreover, some users may "sleep" at some slot due to low battery or busyness. That is, users arrive at and depart from the crowdsensing system dynamically. Therefore, the dynamics of users invalidate the offline incentive mechanism in the existing works.

Technical Challenges There are three technical coupled problems studied in this chapter. For a time slot, (1) How does the platform make pricing decisions on rewards for the coming tasks? (2) Then how do users select tasks to perform under given constraints? (3) After being aware of users' interest of tasks, how does the platform select users to cover all the tasks in this time slot? When solving the first technical problem, there are no prior knowledge of requesters' demands and future information about users' availability and current interest information. It pushes the pricing decision problem into an online decision problem. When users make the decision to reveal their interest of tasks, they are constrained to calculate the optimal subset of tasks as their interests since users have different computing capabilities in their own hand-held mobile phones. This challenge is referred to as bounded rationality [15]. After receiving the interest information about tasks from users, the platform confronts the worker selection problem. The challenge of the worker selection problem is twofold: (1) users' sensing costs are their private information and inaccessible to the platform; (2) the worker selection problem is NP-hard (as proved in Sect. 2.3.1). While we aim to guarantee the fairness of users to be selected as workers in order to ensure users' long-term participation, the already hard problem with the new constraint becomes more challenging. Although the existing works [9, 12, 13, 16] use auction theory to truthfully elicit users' private sensing cost and pay them VCG payment, the auction-based incentive mechanism is not suitable for the scenario. This is because we focus on the platform-centric crowdsensing scenario [10, 11] where the platform has to make pricing decisions and allocate tasks as soon as possible at the beginning of each time slot. In our focused setting, the platform first strategizes the rewards for the tasks received at the beginning of time slot t and notifies the active users with task-reward information. Each user will calculate a set of tasks as his/her interest and report to the platform. The platform finally calculates a set of users to cover and perform the current tasks at the end of time slot t. We find that the strategic workflow at each time slot coincides with sequential interactions in the three-stage Stackelberg game. In contrast, an auction requires users to have an intrinsic value about tasks and thus assumes that users will automatically bid for tasks. The strong assumption requires users' proactive interaction with the platform. We design an incentive mechanism where users can join the crowdsensing in a passive or quasi-passive manner.

2.1.3 Contributions

In this chapter, we propose a three-stage Stackelberg-based incentive mechanism for tackling the evolving task demands and the dynamics of users in crowdsensing. The three technical problems described above in each time slot are modeled by a three-stage Stackelberg game. In the first stage, the platform makes online pricing decisions using our proposed online learning algorithm, which integrates Zinkevich's online gradient learning technique in online convex optimization and drift-plus-penalty technique from Lyapunov opportunistic optimization. In the second stage, the users select their interest of tasks based on our proposed a fully polynomial time approximation scheme algorithm (FPTAS). In the last stage, the platform solves the online worker selection problem which considers the fairness constraint and selects users according to the derived solution. We summarize the contributions of this chapter as follows:

- We consider a time-varying scenario for crowdsensing with the uncertainty of task demands and dynamics of users. We uniformly address the complicated interaction among requesters, users and the platform. We propose an incentive mechanism that guarantees fairness for users to be selected as workers in order to promote users' long-term participation, satisfies users' bounded rationality to reveal interest of tasks, and ensures long-term profits for the platform in order to maintain the sustainability of the crowdsensing system.
- We design an online pricing algorithm by integrating Zinkevich's online gradient learning approach and the drift-plus-penalty technique for the first stage. It allows the platform to make a tradeoff between utility maximization and profit-seeking. For the second stage, we capture the variety of users' computing capabilities and propose an FPTAS for them to report their interest information based on their own computing capability. In the final stage, we design an online worker selection algorithm by combining the approximation algorithm of set multi-cover with Lyapunov optimization. It enables the platform to comprise the guaranteeing approximation with fairness sacrifice.
- We evaluate our proposed incentive mechanism by numerical simulations. The results are presented to verify the fairness for users to be selected as workers and the profits for the platform from the long-term perspective.

2.1.4 Related Work

The studies of crowdsensing mainly include task assignment [6, 17, 18] and incentive mechanism design which is mostly related to our work. In this subsection, we extensively conduct a literature review on incentive mechanism design for crowdsensing from the following aspects:

Stackelberg-Based Incentive Mechanism Stackelberg game models the sequential interaction between the platform and users, in which the platform as a leader first makes a pricing decision, and then users as followers strategize on sensing effort on tasks based on the priced rewards. As a staged game, it is commonly used to model the incentive mechanism design problem. Yang et al. [10, 11] address the incentive scenario where the platform as a leader first makes a pricing decision, and then users as followers strategize on sensing effort (e.g., working time) on tasks based on the priced rewards. As a complement, Nie et al. [19] design an incentive mechanism using a two-stage game by taking social network effects into account. Similar to [19], Cheung et al. [20] investigate the impact of a social network effect as well as user diversity on incentive mechanism by formulating it as a two-stage decision problem (essentially the Stackelberg framework). Xiao et al. [21] model a secure mobile crowdsensing game using the Stackelberg game, and calculate the Stackelberg equilibrium for users to choose optimal decision (i.e., the actual sensing accuracy level and the sensing effort) and for the platform to determine the best payment policy. Those incentive mechanisms only work for one task and assume that users fully participate in crowdsensing. In this chapter, we focus on a common scenario where there are multiple requesters generating tasks at a time slot, and we allow users to "sleep" at some moments.

Auction-Based Incentive Mechanism Auction theory is used to handle asymmetric information (i.e., private sensing cost) when the platform attempts to incentivize users to perform sensing tasks. [10, 11] and [9] propose a truthful reverse auction to incentivize users to bid for tasks so that the platform calculates winning users and VCG payments for them. Duan et al. [16] combines users' preferences with the truthful reverse auction. Wei et al. [13] and Zhang et al. [22] enable the complicated interaction between multiple requesters and multiple users using double auction without considering the quality of collected data. Thus, [14] proposes a double auction-based incentive mechanism with a data aggregation mechanism for guaranteeing quality. Wen et al. [12] proposes a quality-driven auction-based incentive mechanism while taking multiple requesters and fair selection for users into account. However, auction-based incentive mechanisms work in a participatory manner. In this chapter, we take multiple requesters and the fairness for users into account and propose a three-stage Stackelberg-based incentive mechanism in a passive or quasi-passive manner.

Other Games-Based Incentive Mechanisms There are some existing works exploring other games like static games with externality (i.e., network effect [23]), cooperative game (e.g., bargaining [24] and contracting [25]), and Markov game [26] to model incentive mechanism in crowdsensing.

Those above works either only consider static scenarios and a single requester, or neglect the fairness for users. In this chapter, we pay attention to jointly address these issues which are not well-investigated in the above works.

2.2 Game Modeling

Before introducing the system model, stating the studied problem, and analyzing the theoretical results, we summarize the main notations in Table 2.1 for ease of exposition.

Table 2.1 Notations used in Chap. 2

Notations		Meanings
Users	$\mathcal{N}, N$	User set with size N
	S_i^t	Subset of task user i is interested in
	c_{ij}^t	Sensing cost for user i to perform task j
	β_i^t	Sensing capacity of user i
	x_i^t	Allocation variable for user i
	$\overline{x}_i$	Time average allocation rate for user i
	D_i	Dropout probability of user i
	η_i	Maximum number of tasks assigned to user i
Platform	p_j^t	Unit payment received from requester j
	$\boldsymbol{p}_t$	Unit payment vector
	r_j^t	Reward priced for task j
	$\boldsymbol{r}_t$	The reward vector
	d_j^t	The number of users interested in task j
	$\boldsymbol{d}_t$	Tasks' interest vector
	$U_t(\boldsymbol{r}_t|\boldsymbol{d}_t)$	Utility of the platform
	B	The platform's pricing budget
Tasks	M	The number of requesters
	$\mathcal{M}_t, \boldsymbol{M}_t$	Task set and its size
	k_j^t	The number of users required to perform task j
	θ_j	The battery expenditure of task j

2.2.1 Task Model

We assume that M requesters subscribe to a crowdsensing platform. At each slot t,[1] a set of requesters post a number of task requests to the platform. Note that each requester can only post at most one task at each time slot. The set of tasks at time slot t is defined as M_t with $|M_t| = M_t$. Each task $j \in M_t$ is associated with a unit payment p_j, which follows an unknown but fixed distribution (depending on the budget of the task requester). For $j \notin M_t$, $p_j = 0$. Note that the requesters can tolerate the delay of tasks that are performed by users for a certain duration.[2] When task $j \in M_t$ is performed, it consumes computational resources and storage in users' devices, thus generating battery expenditure θ_j.

2.2.2 Platform Model

At the beginning of slot t, the platform receives a set M_t of tasks from requesters. Based on the unit payments $\boldsymbol{p}_t = (p_1^t, p_2^t, \ldots, p_M^t)$ received at slot t, the platform makes pricing decision $\boldsymbol{r}_t = (r_1^t, r_2^t, \ldots, r_M^t)$ which can be interpreted as rewards to attract users' participation. After publicizing pricing decision $\boldsymbol{r}^t$ to users, the platform will receive an interest information $\boldsymbol{d}_t$ from users, which indicates that $d_j^t \in \boldsymbol{d}_t$ users are interested in task j ($\forall j \in M_t$). Note that $\boldsymbol{d}_t$ is unknown to the platform until the users reveal the interest information to the platform. After receiving users' interest information about tasks, the platform achieves a utility $U_t(\boldsymbol{r}_t|\boldsymbol{d}_t)$. We assume that $U_t(\boldsymbol{r}_t|\boldsymbol{d}_t)$ is continuous and concave over $\boldsymbol{r}_t$. A common utility function is [10, 11, 27]:

$$U_t(\boldsymbol{r}_t|\boldsymbol{d}_t) = \sum_{j \in M_t} \log(1 + r_j^t d_j^t), \tag{2.1}$$

where $r_j^t d_j^t$ can be interpreted as platform's maximal investment on task j and $\log(1 + r_j^t d_j^t)$ captures diminishing return property of maximal investment on task j. We have the following lemma to characterize the utility function $U_t(\boldsymbol{r}_t|\boldsymbol{d}_t)$ the platform aims to maximize.

Lemma 2.1 $U_t(\boldsymbol{r}_t|\boldsymbol{d}_t)$ is concave in $\boldsymbol{r}_t$.

---

[1] Consider time horizon of T rounds in this chapter for convenience. However, our approach can be extended to any time.

[2] The impact of delay will be studied in future work

Proof When d_t is given a priori, the utility function $U_t(r_t|d_t)$ is concave due to the concavity of $\log(\cdot)$. If not this case, d_t is dependent on r_t. For task j, we denote by $d_j^t(r_j^t)$ the reward-participation function. It is reasonable to assume that $d_j^t(r_j^t)$ is concave with diminishing return, as the relationship between reward and participation is proved as diminishing return relation in [27]. It means that users' participation increases with reward but the marginal return decreases. By checking the first-order and second-order conditions, we can conclude that the utility $\log(1 + r_j^t d_j^t(r_j^t))$ in task j is concave in r_j^t. Extend the analysis to Hessian matrix [28] of $U_t(r_t|d_t)$, we complete the proof.   □

The platform aims to maximize its utility in the long term. The feasible r_t should satisfy:

$$\mathcal{R} = \{r_t | r_t \cdot \mathbf{1}^\mathsf{T} \leq B\}, \tag{2.2}$$

where B is a maximal rewarding budget. Moreover, the platform aims to make a profit in the long run and has time average pricing constraint:

$$\overline{p_j} = \lim_{t \to \infty} \frac{\sum_{\tau=0}^{t} p_j^\tau}{t} \geq \lim_{t \to \infty} \frac{\sum_{\tau=0}^{t} r_j^\tau}{t} = \overline{r}_j. \tag{2.3}$$

For ease of exposition, we define $\overline{p} = (\overline{p_1}, \overline{p_2}, \dots, \overline{p_j}, \dots, \overline{p_M})$ and $\overline{r} = \cdot(\overline{r_1}, \overline{r_2}, \dots, \overline{r_j}, \dots, \overline{r_M})$.

Besides, the platform has to complete worker selection after receiving the user's interest information. Moreover, the platform selects k_j^t users to perform task j ($\forall j \in \mathcal{M}_t$) in order to improve the accuracy of the sensed data and guarantee the quality of the task. Obviously, k_j^t is bounded by d_j^t (i.e., $k_j^t \leq d_j^t$) because the number of available users to be selected for task j is no more than the number of users who are interested in the task j. Note that k_j^t is predefined by the platform after knowing d_j^t.

2.2.3 User Model

Suppose that there are N mobile users who register in the crowdsensing system, denoted by $\mathcal{N} = \{1, 2, \dots, N\}$. The users roam around different points of interest (POI) over the time slot. The users' locations at each time slot are determined by the users' mobility model. However, user i has a cost (denoted by c_{ij}^t) to perform sensing task j at time slot t, such as transportation fee from the location of user i to the location of task j. Moreover, users have sensing capacities due to the limited remaining battery level. We denote the sensing capacity by β_i^t for user i at time slot t. Note that users' sensing capacities are their private information and are unknown to the platform. We assume that β_i^t is i.i.d over time for user i ($\forall i \in \mathcal{N}$). Based on the publicized rewards r_t and sensing constraints, user i determines a subset of tasks $\mathcal{S}_i^t \subseteq \mathcal{M}_t$. Users inform the determined tasks $\mathcal{S}_i^t$ to the platform. The platform will be aware of d_j^t after users reveal the information of interests to the platform, i.e., $d_j^t = \sum_{i \in \mathcal{N}} \mathbf{1}_{\{j \in S_i^t\}}$, where $\mathbf{1}_{\{\cdot\}}$ equals to 1 when the underlying condition is true.

The crowdsensing system evolves over time such that tasks are allocated to the selected users for completion at each time slot. However, the enthusiasm of the unselected users will fade away if they are not selected to be workers in the long term. Hence, those users drop out of the crowdsensing system. To this end, the platform should ensure the following constraint to guarantee users' long-term participation [29]:

$$\liminf_{t \to \infty} \frac{\sum_{\tau=0}^{t} x_i^t}{t} = \bar{x}_i \geq D_i, \forall i \in \mathcal{N}, \tag{2.4}$$

where $x_i^t \in \{0, 1\}$ is a binary variable to indicate whether user i is selected at time slot t. $\bar{x}_i$ is time average allocation[3] rate for user i and D_i is the threshold of participation rate below which user i will drop out of the crowdsensing system. The constraint in Eq. (2.4) implies that the rate for user i to be selected as a worker must be greater than D_i so that user i will continuously participate in crowdsensing. Based on the above description, we have the following definition to characterize the fairness of the incentive mechanism.

Definition 2.1 (Fairness) An incentive mechanism is said to be fair if and only if the inequality in Eq. (2.4) holds for all user $i \in \mathcal{N}$ and the tasks assigned to each user i are no more than η_i.

The first part (i.e., Eq. (2.4)) of fairness definition follows by Gao et al. [29] and Li et al. [30] which focuses on guaranteeing users' long-term participation and the second part ensures that more users have the opportunity to take part in crowdsensing such that users' extensive participation is guaranteed (i.e., cardinality restrictions). Note that we assume that there are no malicious users in the crowdsensing system and we pay attention to the long-term effect of the designed incentive mechanism. Note that the requesters are not strategic in the proposed incentive mechanism and our incentive mechanism can scale with the number of requesters at each time slot due to the efficiency of algorithms we design (i.e., polynomial time efficiency).

2.2.4   Problem Statement

We use a three-stage Stackelberg game to model the problems studied in this chapter. In each stage, the platform and users have to address different sub-problems and make different decisions. The formulation is:

---

[3] We use selection and allocation interchangeably in this chapter.

- **Stage I:** At the beginning of each time slot t, the platform makes optimal pricing decision r_t^* to maximize the expectation of utility in the time average by solving the stochastic programming:

$$\max \ \liminf_{t \to \infty} \frac{\sum_{\tau=0}^{t} \mathbb{E}[U_t(\mathbf{r}_t|\mathbf{d}_t)]}{t}$$

$$(P1) \qquad s.t. \quad \begin{cases} \bar{\mathbf{r}} \leq \bar{\mathbf{p}}, \\ r_j^t \geq 0, \\ (\mathbf{r}_t \in \mathcal{R}, \ j \in \mathcal{M}_t). \end{cases}$$

The first constraint guarantees the platform's long-term profit and the second one ensures the priced reward is non-negative. Once the platform makes the pricing decision, it will inform to users about the rewards. Note that the current utility $U_t(\mathbf{r}_t|\mathbf{d}_t)$ is unknown until users reveal their interest information in Stage II.

- **Stage II:** When informed the rewards, users determine which tasks they are interested in performing. We model the interests determination problem using the bounded knapsack problem. Taking user i as an example, the formulation is shown as follows:

$$\max \ \sum_{j \in \mathcal{M}_t} r_j^t y_{ij}^t$$

$$(P2) \qquad s.t. \quad \begin{cases} \sum_{j \in \mathcal{M}_t} \theta_j y_{ij}^t \leq \beta_i^t, \\ r_j^t y_{ij}^t \geq c_{ij}^t, \\ \sum_{j \in \mathcal{M}_t} y_{ij}^t \leq \eta_i, \\ (y_{ij}^t \in \{0, 1\}, \ j \in \mathcal{M}_t). \end{cases}$$

The objective is total rewards for user i. The first constraint ensures that the total battery consumed by the tasks users select (i.e., $\sum_{j \in \mathcal{M}_t} \theta_j y_{ij}^t$) is no more than their current sensing capacity (i.e., β_i^t). The second constraint implies that users are only interested in the tasks whose rewards are greater than their current sensing costs. The third constraint indicates that the number of tasks user i selects is less than η_i due to the fairness definition (Definition 2.1). After solving the problem (P2), user i can construct the interest of tasks $\mathcal{S}_i^t = \{j | y_{ij}^t = 1\}$. Users report their interests of tasks to the platform. After being selected by the platform, they will be paid according to the rewards priced at the Stage I when they complete the sensing tasks.

- **Stage III:** Being aware of users' interests of tasks, the platform selects users to perform their interests of tasks subject to tasks' restrictions and users' long-term participation constraints. The objective is to minimize the total rewarding costs

on time average. We formulate the worker selection problem as the following
stochastic programming:

$$\min \ \limsup_{t \to \infty} \frac{\sum_{\tau=0}^{t} \sum_{i \in \mathcal{N}} R_i^t x_i^t}{t}$$

$$(P3)$$
$$s.t. \quad \begin{cases} \sum_{S_i^t : j \in S_i^t} x_i^t \geq k_j^t, \\ \liminf_{t \to \infty} \frac{\sum_{\tau=0}^{t} x_i^t}{t} = \overline{x}_i \geq D_i, \\ (x_i^t \in \{0, 1\}, i \in \mathcal{N}, j \in M_t), \end{cases}$$

where R_i^t is the rewarding cost for user i to perform S_i^t (i.e., $R_i^t = \sum_{j \in S_i^t} r_j^t$).
The first constraint enables a task to be performed by multiple users such that
the quality of the task is guaranteed. The latter constraint ensures the long-term
participation of users.

The problems at each stage are so coupled that it is impossible to independently
optimize them. This is because: (1) The pricing decision at the Stage I affects the
result of problem (P2) but requires the information determined from problem (P2).
(2) The derivation of problem (P2) at the Stage II affects the result2 of problem2
(P3) and (P1) but requires the information of rewards priced at problem (P1).
(3) Finally, addressing problem (P3) at the Stage III requires the interest of tasks
determined from problem (P2) but affects the problem (P2) in turn. The above
description indicates that the platform strategizes in Stage I and Stage II while the
users are strategic in Stage II. Although the information is incomplete at each stage,
it is possible to derive the Stackelberg Equilibrium (SE) using backward induction
analysis [31–33].

2.2.5   Edge-Cloud Implementation

Since the incentive mechanism is compute-intensive, the cloud architecture brings
large latency for the platform to coordinate the interactions between requesters
and users. Luckily, the promising Mobile Edge Computing (MEC) brings the
opportunity to accommodate the challenge of traditional cloud-based crowdsensing.
In edge-cloud-based crowdsensing, the platform offloads the computing tasks of
incentive mechanism (pricing rewards, determining interest of tasks and selecting
workers) to edge servers in the network periphery which is proximal to end-users.
The platform in the cloud server only receives the service requests from subscribers
and forwards the sensing tasks to corresponding edge servers according to the POI
of tasks. Once accepting tasks from the cloud server, the edge servers run the three-
stage Stackelberg-based incentive mechanism to recruit participants from mobile
users to perform tasks. When the edge servers collect sensing data from participants,
they send the sensing results with data aggregate to the platform and the platform
serves subscribers with the sensing results. Following by the edge-cloud architecture

[34, 35], we believe that the performance of our incentive mechanism is efficient. One of the advantages of edge-cloud architecture for crowdsensing is to preserve users' privacy such that the security is improved.

2.3 Detailed Design

2.3.1 Stage III: Online Worker Selection

In this subsection, we analyze the solution to problem (P3). We first present the following lemmas to characterize problem (P3).

Lemma 2.2 *Problem (P3) is NP-hard.*

Proof Actually, (P3) can be regarded as a Bounded Set Cover problem. We can reduce the original Set Multiple Cover (SMC) problem to (P3) with attaching additional incentive constraint and long-term participation constraint (the second constraint and third constraint in (P3), respectively). The ground set is the set of tasks at time slot t (i.e., $\mathcal{M}_t$). The collection of subsets consists of users' interests of tasks. Due to the NP-hardness of SMC [36], this lemma obviously holds. □

Lemma 2.3 *If there exists a task $j^\dagger \in \mathcal{M}_t$ such that $\sum_{i \in \mathcal{N}} 1_{\{j^\dagger \in \mathcal{S}_i^t\}} < k_{j^\dagger}^t$, then the problem (P3) has no solution.*

Proof According to contradiction, we assume that the condition holds while problem (P3) has a solution $\boldsymbol{x}^t = (x_1^{*t}, x_2^{*t}, \ldots, x_N^{*t})$. The solution $\boldsymbol{x}^t$ satisfies the cover constraint $\sum_{\mathcal{S}_i^{*j} : j \in \mathcal{S}_i^{*j}} x_i^t = k_j^t$. However, the cover constraint violates the premise $\sum_{i \in \mathcal{N}} 1_{\{j^\dagger \in \mathcal{S}_i^t\}} < k_j^t$. The contradiction leads to the correctness of this lemma. □

To guarantee users' long-term participation, it suffices to ensure fairness that each user has an opportunity to be selected as a worker over time slot. The fairness is guaranteed if the constraint in Eq. (2.4) is satisfied in the long term. We use the virtual queue technique [37] to accommodate the time-average constraint in Eq. (2.4). The key idea is to ensure the constraint in Eq. (2.4) is equivalently to stabilize the corresponding virtual queues when the platform selects workers. To this end, we define a virtual queue for user $i \in \mathcal{N}$ whose backlog (queue length) at time slot t is denoted by $Q_i(t)$. The backlog $Q_i(t)$ evolves over time slot based on the following dynamic equation which is transformed from the constraint in Eq. (2.4):

$$Q_i(t+1) = \max\{Q_i(t) - x_i^t + D_i, 0\}, \tag{2.5}$$

The backlog vector is $\boldsymbol{Q}(t) = (Q_1(t), Q_2(t), \ldots, Q_N(t))$. The backlog of the virtual queue can be interpreted as a virtual allocating request or allocating debt for a user. The larger $Q_i(t)$ is, the further deviation from the constraint in Eq. (2.4) is. A worker selection policy makes the decision of selecting users to stabilize all users' virtual queues such that the constraint in Eq. (2.4) is satisfied for each user in the long term. We formally define queue stability as follows:

Definition 2.2 A queue system $Q(t)$ is strongly stable if and only if the following condition holds:

$$\limsup_{t \to \infty} \sum_{\tau=0}^{t} \mathbb{E}[\sum_{i=1}^{N} Q_i(\tau)] < \infty. \tag{2.6}$$

We define a scalar function to currently measure the total size of virtual queues (like the norm for vectors):

$$L(Q(t)) = \frac{1}{2} \sum_{i=1}^{N} \omega_i Q_i^2(t), \tag{2.7}$$

where ω_i is the weight to evaluate the importance of $Q_i(t)$. We set $\omega_i = 1$ to equally guarantee fairness among all users in this chapter. The function $L(Q(t)) \geq 0$ in Eq. (2.7) is referred to as Lyapunov function [37] which has three following properties: (1) $L(Q(t)) = 0$ indicates that all virtual queues are empty at time slot t; (2) smaller $L(Q(t))$ implies that the backlogs of all virtual queues are smaller at time slot t; (3) larger $L(Q(t))$ represents that at least one virtual queue has a larger backlog. While we develop an online worker selection algorithm to control the bound of the Lyapunov function over time slot, the virtual queues are desired to be stabilized. To this end, the online worker selection algorithm is proposed to minimize the change of the Lyapunov function for two consecutive time slots. The change of Lyapunov function for two consecutive time slots is defined as Lyapunov drift which is mathematically represented as $\Delta(Q(t)) \triangleq \mathbb{E}[L(Q(t+1)) - L(Q(t))|Q(t)]$. While the platform selects users to guarantee long-term participation constraints by minimizing the Lyapunov drift at each time slot, it has another objective, such as selecting users with minimum sensing costs such that the incentive rewards are paid as less as possible to save the platform's expenditure. However, users' sensing costs are private information and inaccessible to the platform. Instead, the platform aims to minimize the rewards based on current pricing decisions while stabilizing $Q(t)$. We have the following lemma to characterize the platform's online worker selection policy.

Lemma 2.4 *For users in N, if their interests of tasks cover all tasks M_t, the platform selects the set S_t^* of users according to the following equation:*

$$S_t^* = \arg\max_{S \in 2^N} \sum_{i \in S} [Q_i(t) - V \sum_{j \in S_i^t} r_j^t], \tag{2.8}$$

where S_t^* should satisfy the cover requirement of tasks $\mathcal{M}_t$ and V is a control parameter tuned by the platform to make a tradeoff between stabilizing virtual queues and minimizing the incentive rewards of selected users.

Proof Eq. (2.8) can be derived by leveraging the drift-plus-penalty technique [37]. To calculate the bound of drift, we have

$$L(\boldsymbol{Q}(t+1)) - L(\boldsymbol{Q}(t)) \tag{2.9a}$$

$$= \frac{1}{2} \sum_{i=1}^{N} Q_i^2(t+1) - \frac{1}{2} \sum_{i=1}^{N} Q_i^2(t) \tag{2.9b}$$

$$\leq \sum_{i=1}^{N} \frac{D_i^2 + (x_i^t)^2}{2} + \sum_{i=1}^{N} Q_i(t)D_i - \sum_{i=1}^{N} Q_i(t)x_i^t \tag{2.9c}$$

$$\leq N + \sum_{i=1}^{N} Q_i(t)D_i - \sum_{i=1}^{N} Q_i(t)x_i^t, \tag{2.9d}$$

where Eq. (2.9c) holds by the fact of $(\max\{Q - x + D, 0\})^2 \leq Q^2 + x^2 + D^2 + 2Q(D - x)$, Eq. (2.9d) holds because D_i and x_i^t lie in the range $[0, 1]$. Taking the conditional expectation on both sides of Eq. (2.9) with the respect to current backlogs $\boldsymbol{Q}(t)$, we have:

$$\Delta(\boldsymbol{Q}(t)) \leq N + \sum_{i=1}^{N} Q_i(t)D_i - \mathbb{E}[\sum_{i=1}^{N} Q_i(t)x_i^t \,|\, \boldsymbol{Q}(t)]. \tag{2.10}$$

The worker selection algorithm only affects the last term of r.h.s in Eq. (2.10). Therefore, the worker selection algorithm can minimize Lyapunov drift by selecting a subset S of users to maximize the current backlogs $\sum_{i \in S} Q_i(t)$ subject to task cover requirements. Additionally, as the worker selection algorithm also focuses on minimizing the incentive rewards to the selected workers, minimizing $\Delta(\boldsymbol{Q}(t)) + V\mathbb{E}[\sum_{i \in S} \sum_{j \in S_i^t} r_j^t \,|\, \boldsymbol{Q}(t)]$ equivalently maximizes $\sum_{i \in S}[Q_i(t) - V \sum_{j \in S_i^t} r_j^t]$. Note that S should cover all tasks by multiple times.   □

Based on the worker selection policy in Lemma 2.4, we design the Online Worker Selection algorithm (Algorithm 2.1). In Algorithm 2.1, we input the pricing decision vector r_t, user set N, task set $\mathcal{M}_t$ at time slot t, the threshold of participation rate D_i for user i, control parameter V, the maximum number k_j^t of users selected to perform task $j \in \mathcal{M}_t$, and the users' interests collection $\{S_1^t, S_2^t, \ldots, S_N^t\}$. In line 1, we define a set C to record the selected users, which is initialized to be an empty set. Besides, we also define a counter variable ζ_j^t with initial value 0 for task j. Before we make the worker selection decision, we first update the virtual queues backlogs $\boldsymbol{Q}(t)$ based on the previous decision and according to Eq. (2.5) (line 2). In lines 3–6, we define other variable C' (candidate user set) with value N and

---

Algorithm 2.1 Online worker selection

Input: r_t, N, M_t, $D_i(\forall i \in N)$, V, $k_j^t(\forall j \in M_t)$ and $\{S_1^t, S_2^t, \ldots, S_N^t\}$
Output: C
1: $C = \emptyset$, define temporary variables $\zeta_j^t = 0, \forall j \in M_t$.
2: At time slot t, update $Q(t)$ according to Eq. (2.5).
3: Let $C' = N$.
4: **if** (P3) with respect to C' has no solution according to Lemma 2.3 **then**
5:    **return** C.
6: **end if**
7: **while** $\cup_{\forall j \in M_t} \zeta_j^t < k_j^t$ **do**
8:    For $i \in C'$, set $\alpha_i = \dfrac{\sum_{j \in S_i^t} r_j^t}{|T_i|}$, where $T_i = \{j | j \in S_i^t \text{ and } \zeta_j^t < k_j^t\}$.
9:    $i^* = \arg\max_{i \in C'} Q_i(t) - V\alpha_i$.
10:   $C = C \cup \{i^*\}$.
11:   $C' = C' - \{i^*\}$.
12:   $\zeta_j^t = \zeta_j^t + 1, \forall j \in S_i^t$.
13: **end while**
14: **return** C.

---

use it to check whether the problem has a solution based on Lemma 2.3. If no solution, the algorithm returns an empty set, which indicates that the platform fails to select proper workers to satisfy the tasks' requirements. Otherwise, the selected user set C is derived in lines 7–14. The key operations are to (1) calculate selection cost α_i for user $i \in C'$ (line 8), (2) label metric $Q_i(t) - V\alpha_i$ for user $i \in C'$, (3) find a user i^* with the maximum metric $Q_i(t) - V\alpha_i$ from $mathcalC'$ (line 9), (4) update variables C, C', and ζ_j^t (line 10–12). After the key operations, we return the obtained worker set C for the platform to select (line 14). The time complexity of Algorithm 2.1 is dominated by the **while** loop in line 7–13, which depends on the condition $\cup_{\forall j \in M_t} \zeta_j^t < k_j^t$. Therefore, the time complexity is $O(|M_t|K)$ at most, where $K = \max_{j \in M_t} k_j^t$.

Note that by subtle adaptation, Algorithm 2.1 can actually be extended to the scenario where some users are sleeping at some slots. Assuming that $N_t \subseteq N$ is the set of online users at time slot t, the platform should update the queue backlogs of users in N_t using Eq. (2.5) and the remaining operations are run as Algorithm 2.1 does. For "sleeping" users (i.e., $N \backslash N_t$), we keep their queue backlogs unchanged (i.e., $Q_i(t) = Q_i(t-1), \forall i \in N \backslash N_t$). This can be interpreted that for "sleeping" users, there are no virtual allocating requests for them or the platform has no allocating debts for them. By doing the adaptation, Algorithm 2.1 can perfectly be extended as the variant to handle with "sleeping" users scenario.

2.3.2 Stage II: Users' Interests of Tasks Disclosure

When notified with the pricing decision r_t for tasks M_t at the Stage I, users have an incentive to disclose their interests of tasks to the platform since they will receive the rewards if they are selected to be workers. Moreover, Algorithm 2.1 at the Stage III guarantees the fairness requirement that each user has an opportunity to be a worker in the long term.

As stated in Sect. 2.2.4, users determine their interests of tasks by solving problem (P2). The problem (P2) is equivalent to the Knapsack Problem where the objects are tasks with battery expenditures as sizes, and profits of the objects are rewards and the knapsack capacity is users' sensing capacities β_i^t ($\forall i \in N$). However, the differences from the conventional Knapsack Problem are twofold: (1) we consider the incentive constraint $(r_j^t y_{ij}^t \geq c_{ij}^t)$; (2) we impose the restriction on the number of tasks that a user can select according to the fairness definition (Definition 2.1). In addition to this diversity, with subtle adaptation, (P2) can be solved by using the classical dynamic programming based on the following dynamic equation:

$$H(j+1, \gamma) = \begin{cases} \min\{H(j, \gamma), \theta_{j+1} + H(j, \gamma - r_{j+1}^t)\}, & \text{if } c_{i,j+1}^t \leq r_{j+1}^t \leq \gamma \\ H(j, \gamma), & \text{otherwise,} \end{cases}$$

(2.11)

where $H(j, \gamma)$ is the total battery expenditures of the selected tasks and γ is the total rewards of the selected tasks, with respect to the task set $\{1, 2, \ldots, j\}$. The objective of (P2) is $\max\{\gamma | H(M_t, \gamma) \leq \beta_i^t\}$ when user i chooses S_i^t. Users can solve the dynamic programming for (P2) within $O(M_t^2 R)$ where $R = \max_{j \in M_t} r_j^t$. This is a pseudo-polynomial time algorithm whose complexity is dominated by the number of tasks M_t at time slot t and maximum reward R. We develop the following approximation algorithm to calculate the solution by relaxing r_j^t to $\tilde{r}_j^t$ with gap $G = \frac{\varepsilon_i R}{M_t}$. Specifically, ε_i is a relaxed factor tuned by user i.

Users' interests of tasks revelation algorithm is shown in Algorithm 2.2, which takes the pricing decision vector r_t, task set M_t, sensing capacity β_i^t for user i at time slot t, the maximum number η_i of the selected tasks, resource expenditure θ_j of performing task j, sensing cost c_{ij}^t for user i to perform task j at time slot t, and relaxed factor ε_i for user i as input. Note that each user i independently runs Algorithm 2.2 on its device to reveal its interest of tasks and reports $\tilde{S}_i^t$ to the platform. In line 1, user i selects the tasks satisfying the incentive constraint $r_j^t y_{ij}^t \geq c_{ij}^t$ from M_t and obtains M_t'. In line 2, user i derives a discount factor G based on its relaxed factor ε_i. The new prices ($\forall j \in M_t'$) are determined using the discount factor G for all tasks in M_t' (line 3–5). In line 6, user i chooses the tasks S_i^t with the maximum potential rewards from M_t' using dynamic programming based on the new prices $\tilde{r}_j^t$ ($\forall j \in M_t'$). If the size of S_i^t is less than η_i, the algorithm directly outputs it (lines 7–9). Otherwise, the algorithm sorts the tasks in increasing order of new prices $\tilde{r}_j^t$ ($\forall j \in M_t'$) and chooses the top η_i tasks to output (line 10–13).

Algorithm 2.2 Interest revelation

Input: $r_t, \mathcal{M}_t, \beta_i^t, \eta_i, \theta_j, c_{ij}^t \;\; \forall j \in \mathcal{M}_t$ and ε_i

Output: $\mathcal{S}_i^t$

1: Select the candidate tasks from $\mathcal{M}_t$ according to incentive constraint $r_j^t y_{ij}^t \geq c_{ij}^t$ ($\forall j \in \mathcal{M}_t$), denote by $\mathcal{M}_t'$.

2: Assert $\varepsilon_i > 0$, set $G = \frac{\varepsilon_i R}{|\mathcal{M}_t'|}$.

3: **for** $j \in \mathcal{M}_t'$ **do**

4:     Relax the rewards, $\tilde{r}_j^t = \lfloor \frac{r_j^t}{G} \rfloor$.

5: **end for**

6: Calculate the interest of tasks $\mathcal{S}_i^t$ over $\mathcal{M}_t'$ using the above dynamic programming (i.e., Eq. (2.11)) based on the relaxed rewards $\tilde{r}_j^t$.

7: **if** $|\mathcal{S}_i^t| \leq \eta_i$ **then**

8:     **return** $\mathcal{S}_i^t$.

9: **else**

10:     Sort all $j \in \mathcal{S}_i^t$ based on $\tilde{r}_j^t$ in a decreasing order π.

11:     Define $\tilde{\mathcal{S}}_i^t = \{\pi_1, \pi_2, \ldots, \pi_{\eta_i}\}$.

12:     **return** $\tilde{\mathcal{S}}_i^t$.

13: **end if**

2.3.3   Stage I: Online Pricing

For online scenario, the platform has to determine the pricing decision $\tilde{r}_t$[4] to maximize $U_t(\tilde{r}_t|\tilde{d}_t)$ at the beginning of time slot t. However, $\tilde{d}_t$ is unknown until users reveal their interests of tasks to the platform in Stage II. To this end, we develop an approach to address the online pricing problem (P1) by integrating the drift-plus-penalty technique in Lyapunov optimization with Zinkevich's online gradient method [38–41] in online convex programming. To guarantee the long-term pricing constraint in Eq. (2.3) when making the pricing decision, we define a budget queue with backlog $Z_j(t)$ for requester j at time slot t. Its dynamics evolve over time by the following equation:

$$Z_j(t+1) = \min\{Z_j(t) + p_j^t - r_j^t, 0\}, \tag{2.12}$$

where $p_j^t = r_j^t = 0$ for $j \notin \mathcal{M}_t$. We have the following lemma to characterize the online pricing decision $\tilde{r}_t$.

Lemma 2.5 *For any tuned parameter $\nu > 0$, the platform determines the pricing decision $\tilde{r}_t$ to maximize an expression $\nu U_t(\tilde{r}_t|\tilde{d}_t) + \widetilde{Z(t)}^\mathsf{T} \tilde{r}_t$ when tasks $\mathcal{M}_t$ is arriving at time slot t, i.e.,*

$$\tilde{r}_t^* = \arg\max_{\tilde{r}_t} \nu U_t(\tilde{r}_t|\tilde{d}_t) + \widetilde{Z(t)}^\mathsf{T} \tilde{r}_t. \tag{2.13}$$

---

[4] From now on, we use notation $\tilde{r}_t$ instead of r_t due to the uncertainty of tasks. In detail, $\tilde{r}_t = (r_{j(1)}^t, r_{j(2)}^t, \ldots)$ where $j(k) \in \mathcal{M}_t, 1 \leq k \leq |\mathcal{M}_t|$. $\tilde{d}_t, \tilde{p}_t$ and $\widetilde{Z(t)}$ have the same meanings.

Proof As usual, we define Lyapunov function as $L(\mathbf{Z}(t)) = \frac{1}{2}\sum_{j\in M_t} Z_j^2(t)$ and Lyapunov drift as $\Delta(\mathbf{Z}(t)) \triangleq \mathbb{E}[L(\mathbf{Z}(t+1)) - L(\mathbf{Z}(t))|\mathbf{Z}(t), j \in M_t \cap M_{t+1}]$. Following by the drift analysis, we have:

$$\Delta(\widetilde{\mathbf{Z}(t)}) \le \frac{(\mathbf{1}^{\mathsf{T}}\widetilde{\mathbf{p}}_t)^2 + B^2}{2} + \widetilde{\mathbf{Z}(t)}^{\mathsf{T}}\widetilde{\mathbf{p}}_t - \widetilde{\mathbf{Z}(t)}^{\mathsf{T}}\widetilde{\mathbf{r}}_t. \tag{2.14}$$

Minimizing $\Delta(\widetilde{\mathbf{Z}(t)})$ is equivalently to minimize $-\widetilde{\mathbf{Z}(t)}^{\mathsf{T}}\widetilde{\mathbf{r}}_t$ by determining $\widetilde{\mathbf{r}}_t$. Plugging the penalty term $-\nu U_t(\widetilde{\mathbf{r}}_t|\widetilde{\mathbf{d}}_t)$ into Eq. (2.14), we prove this lemma. □

Although Lemma 2.5 points out the policy to online pricing problem for the platform, it is still intractable because $U_t(\widetilde{\mathbf{r}}_t|\widetilde{\mathbf{d}}_t)$ is unknown a prior. Fortunately, the fact that $U_t(\widetilde{\mathbf{r}}_t|\widetilde{\mathbf{d}}_t)$ is a concave function allows us to design an online pricing algorithm using Zinkevich's online gradient method in online convex programming. Zinkevich's online gradient method involves the following gradient update equation:

$$\widetilde{\mathbf{r}_{t+1}} = \text{Proj}_{\mathcal{R}}\{\widetilde{\mathbf{r}}_t + \eta\nabla_r U_t(\widetilde{\mathbf{r}}_t|\widetilde{\mathbf{d}}_t)\}, \tag{2.15}$$

where $\text{Proj}_{\mathcal{R}}$ is a projection operator which maps the resulted gradient into a feasible reward set $\mathcal{R}$ and η is the learning rate. Therefore, we can regard Zinkevich's online gradient method as a kind of online learning which learns the policy from the gradient at the previous time slot. When maximizing $U_t(\widetilde{\mathbf{r}}_t|\widetilde{\mathbf{d}}_t)$ according to the update rule in Eq. (2.15), we need to keep the budget queues as stable as possible. Hence, we need to combine Zinkevich's online gradient method with the drift-plus-penalty technique to design an online pricing algorithm. We have the following lemma to show the combination:

Lemma 2.6 *The platform makes pricing decisions at time slot t according to the following rule:*

$$\widetilde{\mathbf{r}}_t = \text{Proj}_{\mathcal{R}}\{\widetilde{\mathbf{r}_{t-1}} + \frac{1}{\nu}\nabla_r U_{t-1}(\widetilde{\mathbf{r}_{t-1}}|\widetilde{\mathbf{d}_{t-1}}) + \frac{1}{\nu^2}\widetilde{\mathbf{Z}(t)}\}. \tag{2.16}$$

Proof Plugging Taylor expansion with respect to $U_t(\widetilde{\mathbf{r}}_t|\widetilde{\mathbf{d}}_t)$ into Eq. (2.13) and eliminating the constant term, we have

$$\widetilde{\mathbf{r}}_t^* = \arg\max_{\widetilde{\mathbf{r}}_t} \nu(\nabla_r U_{t-1}(\widetilde{\mathbf{r}_{t-1}}|\widetilde{\mathbf{d}_{t-1}}))^{\mathsf{T}}(\widetilde{\mathbf{r}}_t - \widetilde{\mathbf{r}_{t-1}})$$
$$-\frac{\nu^2}{2}||\widetilde{\mathbf{r}}_t - \widetilde{\mathbf{r}_{t-1}}||^2 + \widetilde{\mathbf{Z}(t)}^{\mathsf{T}}\widetilde{\mathbf{r}}_t. \tag{2.17}$$

By converting the definition of projection operator $min_{\widetilde{\mathbf{r}}_t\in\mathcal{R}}||\widetilde{\mathbf{r}}_t - (\widetilde{\mathbf{r}_{t-1}} + \frac{1}{\nu}\nabla_r U_{t-1}(\widetilde{\mathbf{r}_{t-1}}|\widetilde{\mathbf{d}_{t-1}}) + \frac{1}{\nu^2}\widetilde{\mathbf{Z}(t)}))||^2$, we can find that it is equivalent to Eq. (2.17). □

Algorithm 2.3 Online pricing

Input: M_t, $\tilde{p}_t$, v
Output: r_j^t ($\forall j \in M_t$)
  1: At time slot $t - 0$, initialize $Z_j(0) = 0$, $\forall j \in \{1, 2, \ldots, M\}$.
  2: At time slot $t > 0$, divide M_t into two subsets as follows:
  3: $\mathcal{A} = \{j | j \in M_t$ and requester j firstly posts task$\}$.
  4: $\mathcal{B} = M_t \backslash \mathcal{A}$
  5: Denote the pricing decision vector for $\mathcal{B}$ by $\tilde{r}_t$.
  6: Derive $\tilde{r}_t$ according to Lemma 2.6.
  7: **for** $j \in \mathcal{A}$ **do**
  8:     set $r_j^t = p_j^t$.
  9: **end for**
 10: Update $Z_j(t)$ according to Eq. (2.12), $\forall j \in M_t$.
 11: **return** r_j^t ($\forall j \in M_t$).

We present the proposed online pricing algorithm in Algorithm 2.3. The algorithm takes the current task set M_t, the tasks' payments $\tilde{p}_t$, and the control parameter v as input. In line 1, the algorithm initializes the budget queue $Z_j(0) = 0$, $\forall j \in \{1, 2, \ldots, M\}$. In lines 2–4, we divide the current tasks M_t into $\mathcal{A}$ and $\mathcal{B}$, where the tasks from $\mathcal{A}$ correspond to the requesters' first publicizing tasks, while the tasks from $\mathcal{B}$ are not the first publicizing tasks. The tasks in $\mathcal{B}$ have historical information $\widetilde{r_{t-1}}$, $\nabla_r U_{t-1}(\widetilde{r_{t-1}}|\boldsymbol{d}_{t-1})$, which can be used to make pricing decisions $\tilde{r}_t$ for the tasks using Lemma 2.6 (line 6). For tasks in $\mathcal{A}$, we assign the price r_j^t to p_j^t (lines 7–9). Finally, we update the budget queues in line 10 and return the pricing decisions in line 11. It is easy to see the time complexity of Algorithm 2.3 is $O(M)$.

2.4 Equilibrium Analysis

In this section, we present an equilibrium analysis of the proposed long-term incentive mechanism to show the strategy outcome of the MCS system under the long-term proposed incentive mechanism.

2.4.1 Strategy Performance in Stage III

Recall that we design an online worker selection algorithm (as shown in Algorithm 2.1) in Stage III. The strategy output by Algorithm 2.1 has the following theoretical performance in Theorem 2.1.

Theorem 2.1 *Algorithm 2.1 achieves the theoretical performance as follows:*

$$\bar{r} = \lim_{T \to \infty} \frac{1}{T} \sum_{t=0}^{T-1} \mathbb{E}[R(t)] \leq O(\log M_t)\mathbb{E}[OPT] + \frac{N}{V}, \tag{2.18}$$

$$\lim_{T \to \infty} \frac{1}{T} \sum_{t=0}^{T-1} \sum_{i=1}^{N} Q_i(t) \leq \frac{N}{\epsilon} + \frac{V(O(\log M_t)\mathbb{E}[OPT] - \bar{r})}{\epsilon}. \tag{2.19}$$

Proof We present the detailed proof in [42].   □

Theorem 2.1 indicates that Algorithm 2.1 can generate a worker set with smaller incentive rewards using larger V. But larger V will increase the backlogs of virtual queues within $O(V)$. It means that Algorithm 2.1 gives priority to optimizing incentive rewards by sacrificing the convergence of long-term incentive constraint in Eq. (2.4) under a larger V scenario. Therefore, Algorithm 2.1 has a performance tradeoff between minimizing incentive rewards and stabilizing virtual queues within $[O(\frac{1}{V}), O(V)]$. When $V \to \infty$, Algorithm 2.1 degenerates a greedy algorithm to output task set multiple cover, which provides a factor $O(\log M_t)$ of approximation.

2.4.2 Strategy Performance in Stage II

The strategy in Stage II corresponds to our proposed Algorithm 2.2, i.e., interest revelation. Before the following analysis, we first give a definition to characterize the approximation algorithm.

Definition 2.3 (FPTAS [36]) Suppose $\mathcal{A}$ is an approximation algorithm. If its time complexity is bounded by the size of the problem instance, $\mathcal{A}$ is referred to as a polynomial time approximation scheme (PTAS). If $\mathcal{A}$ is PTAS and its time complexity is also dependent on a relaxed factor ε which controls the approximation factor, then $\mathcal{A}$ is called FPTAS.

We present the properties of Algorithm 2.2 as follows and omit the proofs as the similar proofs can be found in the Lemma 8.3 and Theorem 8.4 of [36].

Lemma 2.7 *The objective of (P2) with respect to the solution from Algorithm 2.2 satisfies:*

$$Obj(\mathcal{S}_i^t) \geq (1 - \varepsilon_i)OPT. \tag{2.20}$$

Lemma 2.8 *Algorithm 2.2 is an FPTAS to problem (P2).*

Lemma 2.7 indicates that the approximation ratio of Algorithm 2.2 is $1 - \varepsilon_i$. Lemma 2.8 implies that the complexity of Algorithm 2.2 is not only dependent on the size of tasks $|\mathcal{M}_t'|$, but also dominated by the relaxed factor ε_i. This is because the complexity of Algorithm 2.2 is $O(|\mathcal{M}'_t|^2 \lfloor \frac{|\mathcal{M}'_t|}{\varepsilon_i} \rfloor)$. Therefore, we can control the granularity of the solution to (P2). If a fine-grained solution is required, a smaller ε_i should be set and vice versa. Note that each user needs to solve (P2) and independently run Algorithm 2.2 to reveal their interest of tasks under current pricing decision r_t. ε_i is differently determined for different users according to their computing capability.

2.4.3 Strategy Performance in Stage I

The strategy in Stage I corresponds to our proposed Algorithm 2.3, i.e., online pricing. The theoretical performance of Algorithm 2.3 is shown in Theorem 2.2.

Theorem 2.2 *Algorithm 2.3 achieves the theoretical performance as follows:*

$$\bar{U} = \lim_{T \to \infty} \frac{1}{T} \sum_{t=0}^{T-1} \mathbb{E}[U_t] \geq \mathbb{E}[OPT] - \frac{\Omega}{\nu}, \tag{2.21}$$

$$\lim_{T \to \infty} \frac{1}{T} \sum_{t=0}^{T-1} \mathbf{Z}(t)^{\mathsf{T}} \mathbf{1} \geq -\frac{\Omega}{\delta} + \frac{\nu(\mathbb{E}[OPT] - \bar{U})}{\delta}. \tag{2.22}$$

Proof We present the detailed proof in [42]. □

The Theorem 2.2 indicates the backlogs of budget queues $\mathbf{Z}(t)$ are negative, which is different from the virtual queues $\mathbf{Q}(t)$ in Theorem 2.1 with the positive backlogs. This is because the update equation (i.e., Eq. (2.12)) for $\mathbf{Z}(t)$ is different from the one (i.e., Eq. (2.5)) for $\mathbf{Q}(t)$.

Based on the previous analyses of the algorithms in our proposed three-stage incentive mechanism, we can see that our proposed algorithms enjoy polynomial time complexity. In detail, the time complexity of the online pricing algorithm in Stage I is $O(|\mathcal{M}_t|)$ where $|\mathcal{M}_t|$ is the number of tasks at time slot t. The time complexity of user's interest revelation algorithm in Stage II is $O(|\mathcal{M}'_t|^2 \lfloor \frac{|\mathcal{M}'_t|}{\varepsilon_i} \rfloor)$ (pseudo-polynomial). The time complexity of online worker selection algorithm in Stage III is $O(|\mathcal{M}_t|K)$ where K is maximum covering number, i.e., $K = \max_{j \in \mathcal{M}_t} k_j^t$. It could be found that these three algorithms can output the sub-optimal solutions within polynomial time, which is tolerable in a real-world online decision system, compared to the exponential time algorithm like brute-force algorithm.

2.5 Performance Evaluation

In this section, we conduct simulations to validate the performance of our proposed incentive mechanism. The proposed algorithms are implemented by Python (for the parts involving LP and MIP, we use the CPLEX Python interface as an implementation tool).

We consider a crowdsensing simulation setting with 1000 users and 1000 requesters (i.e., $N = 1000$ and $M = 1000$). We focus on the crowdsensing interactions with 1000 slots (i.e., $T = 1000$). At each time slot, there are 30 requesters posting the task demands to the platform and 100 online users at least (i.e., $|\mathcal{M}_t| \geq 30$ and $|\mathcal{N}_t| \geq 100$). For each user i, we set the dropout probability as $D_i \sim U(0.01, 0.11)$ where $U(\cdot, \cdot)$ is the uniform distribution. The payment brought by requester j follows $p_j^t \sim U(0, 1000)$. Similar simulation settings can be found in [11, 29]. The detailed settings will be described in the following section. We run the simulations on a computer with the setting: Intel(R) Core(TM) i7-6700 CPU @3.4 Hz processors with 4 cores, 32 GB RAM and 1 TB disk space.

2.5.1 Evaluation for Stage III

To solely evaluate the performance of Algorithm 2.1, we assume that the pricing decision r_t is given over different time slots. For each time slot, each requester generates a task by certain distribution. Users reveal their interests of tasks over time slot. The number of users selected to perform tasks is randomly sampled (i.e., $k_j^t \sim U(1, d_j^t), \forall j \in \mathcal{M}_t$). Therefore, we have a set multicover instance. To derive the optimum as a benchmark in CPLEX, we assume the total number of tasks is bounded by 30. We implement two other benchmarks: one is randomized rounding and the other one is a greedy algorithm. Note that they are the typical approximation algorithms to set multicover problem which both provide a factor of $O(\log(|\mathcal{M}_t|))$ performance. The benchmarks do not necessarily guarantee users' long-term participation. To evaluate the performance of Algorithm 2.1, we consider two situations where $V = 20$ and $V = 100$ for Algorithm 2.1.

Figure 2.1 shows that the rewarding costs of Algorithm 2.1 increase over time but incline to be stable (i.e., the lines with legends $V = 20$ and $V = 100$). This is because Algorithm 2.1 sacrifices cost to guarantee the selection rate of the user with higher rewarding cost so that the long-term participation constraint is satisfied. If V is larger, Algorithm 2.1 gives priority to minimizing rewarding cost but requires larger queues backlogs as shown in Fig. 2.2. The numerical results in Figs. 2.1 and 2.2 also verify the exactness of Theorem 2.1. Although Algorithm 2.1 results in a larger rewarding cost in worker selection compared with the benchmarks, Algorithm 2.1 can converge to the stable upper bound (i.e., $O(\log \boldsymbol{M}_t)$). While the benchmark neglects to incentivize users' long-term participation, our proposed Algorithm 2.1 can compromise some rewarding cost to guarantee the required

Fig. 2.1 The performance comparison for Algorithm 2.1 with benchmarks on rewarding cost

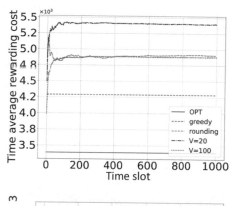

Fig. 2.2 The performance of the total queue backlogs of $Q(t)$ in time average under $V = 20$ and $V = 100$

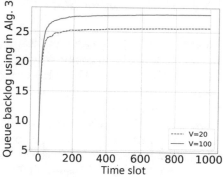

Fig. 2.3 The performance of selection rates for all users under $V = 20$ and $V = 100$

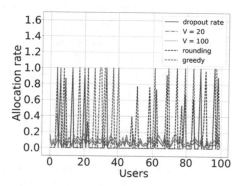

selection rates for users as shown in Fig. 2.3. The dashed line (allocation rate of the benchmark) is below the solid line (users' dropout rate) in some users' id as demonstrated in Fig. 2.3. This result verifies the fairness guaranteeing property of our proposed algorithm.

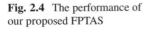

Fig. 2.4 The performance of our proposed FPTAS

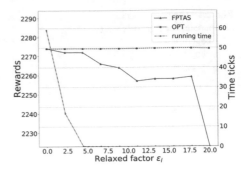

2.5.2 Evaluation for Stage II

We evaluate the performance of our proposed FPTAS (i.e., Algorithm 2.2). The number of tasks is randomly generated and posted to a user i. We set $\beta_i^t \sim U(50, 100)$, $r_j^t \sim U(1, 100)$, and $\theta_j \sim U(1, 5)$, $\forall j \in \mathcal{M}_t$. We set a small c_{ij}^t and η_i greater than $|\mathcal{M}_t|$ such that user i is fully incentivized to select all the coming tasks. ε_i is ranged in $[0, 20]$.

We run the FPTAS to derive the total rewards and running time for user i. As a benchmark, we calculate the optimum (i.e., OPT) using pure dynamic programming. The results are shown in Fig. 2.4. The smaller ε_i is, the more proximal to OPT our FPTAS is. However, FPTAS with smaller ε_i results in higher running time. Thus, it presents a utility-delay tradeoff in terms of calculating the sub-optimal solution. In reality, what relaxed factor should be chosen is determined by the computing capability of users' mobile phones. If the remaining time until the current time slot is smaller, users with lower computing capability could choose a larger relaxed factor ε, or choose the smaller one otherwise.

2.5.3 Evaluation for Stage I

To solely evaluate the performance of Algorithm 2.3, we set default parameters as $M = 100$, $B = 10000$ and $d_j^t \sim U(1000, 10000)$.

Figure 2.5 illustrates that the time-average utility first increases sharply (because Algorithm 2.3 assigns the prices of first arriving tasks to be their payments without making any profits) and gradually converges to the stable optimum. Note that the changing trend of time-average utility fluctuates to the convergence, as it results from three reasons: (1) Algorithm 2.3 learns to converge to the stable optimum using the gradient information it accumulates; (2) The tasks arrive irregularly with different payments; (3) Algorithm 2.3 must maintain the stability of budget queues $\widetilde{Z}(t)$ while optimizing utility. When the platform makes pricing decisions in Algorithm 2.3 using larger ν, the time-average utility first converges. However, the backlogs under larger ν situations are higher than the ones of smaller ν as

Fig. 2.5 The performance comparison for Algorithm 2.3 on time average utility under $v = 10$ and $v = 50$

Fig. 2.6 The performance of the total queue backlogs of $\widehat{Z}(t)$ in time average under $v = 10$ and $v = 50$

Fig. 2.7 The performance of pricing decisions for all tasks under $v = 10$ and $v = 50$

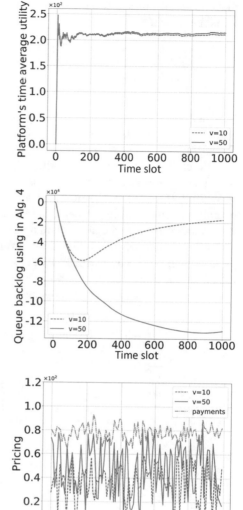

demonstrated in Fig. 2.6. This means that the platform makes a tradeoff between achieving the optimal utility and stabilizing the budget queues when making pricing decisions, which coincides with the conclusion of Theorem 2.2. Figure 2.7 shows that all the rewards pricing for tasks in time average are lower than the payments tasks bring but the time-average rewards of larger v are priced higher than the smaller one due to the nature of utility maximization under larger v situation. Figure 2.7 also verifies that our proposed Algorithm 2.3 enables the platform to make long-term profits when the platform makes pricing decisions to incentivize users' participation.

2.5.4 Evaluation on Trace

We evaluate the performance of the proposed incentive scheme over a synthetic trace within 1000 time slots. We set $M = 1000$, $|\mathcal{M}_t| = 50$, $p_j^t \sim U(0, 1000)$, $\forall j \in \mathcal{M}_t$, $N = 100$, $|\mathcal{N}_t| \geq 50$, $D_i \sim U(0.01, 0.11)$, and $\varepsilon_i \sim U(0, 20)$, $\forall i \in \mathcal{N}$.

At each time slot, we use Algorithm 2.3 to make pricing decisions. After users reveal interests by Algorithm 2.2, we use Algorithm 2.1 to select users. We consider $V = 20, 100, 1000$ for Algorithm 2.1 and $\nu = 10, 20, 210$ for Algorithm 2.3. The overall performances are presented in Figs. 2.8, 2.9, and 2.10. As shown in Fig. 2.8, the rewards offered by the platform increase with ν since the platform will set higher rewards for tasks when ν is larger according to our theoretical analysis. But the platform will select a user with smaller rewards under a larger V situation. However, the queue backlog using in Algorithm 2.1 increases with V as illustrated in Fig. 2.9 and the queue backlogs using in Algorithm 2.3 decreases with ν as demonstrated in Fig. 2.10.

Fig. 2.8 The performance comparison on trace under Algorithm 2.1 with $V = 20, 100, 1000$ and Algorithm 2.3 with $\nu = 10, 20, 210$

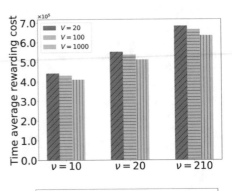

Fig. 2.9 The queue backlog using in Algorithm 2.1 with $V = 20, 100, 1000$ during 1000 slots when evaluating on trace

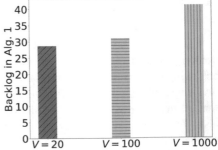

Fig. 2.10 The queue backlog using in Algorithm 2.3 with $\nu = 10, 20, 210$ during 1000 slots when evaluating on trace

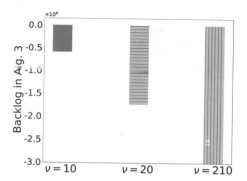

Fig. 2.11 The average profit performance comparison with baselines on the trace

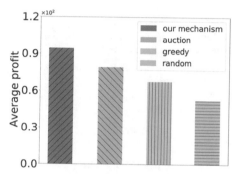

Moreover, we compare our proposed mechanism with the following baselines in terms of average profit, allocation rate and running time. The baselines include,

- `Auction` [9, 13, 16]: This baseline uses an auction to select workers and determine the corresponding rewards based on the VCG mechanism.
- `Greedy`: This baseline greedily selects workers based on their bids and randomly determines the task's reward based on the sensing cost and payment.
- `Random`: This baseline randomly selects workers and randomly determines the task's reward based on the sensing cost and payment.

Specifically, we calculate the average profit by dividing the total time slots and total tasks. The comparisons are demonstrated in Figs. 2.11, 2.12, and 2.13. In Fig. 2.11, we can see that our proposed mechanism can make more profits compared to the baselines from the long-term perspective. In Fig. 2.12, it shows that our proposed mechanism can keep the fair allocation rate (red line) above the dropout probability (blue line) while other baselines fail to keep a fair allocation rate. In Fig. 2.13, it shows that our proposed mechanism can finish the decisions within 25ms at a time slot similar to `Greedy` and `Random`. But the `Auction` costly consumes 28s to finish one-round decisions. Totally, the results verify the superiority and outperformance of our proposed three-stage incentive mechanism.

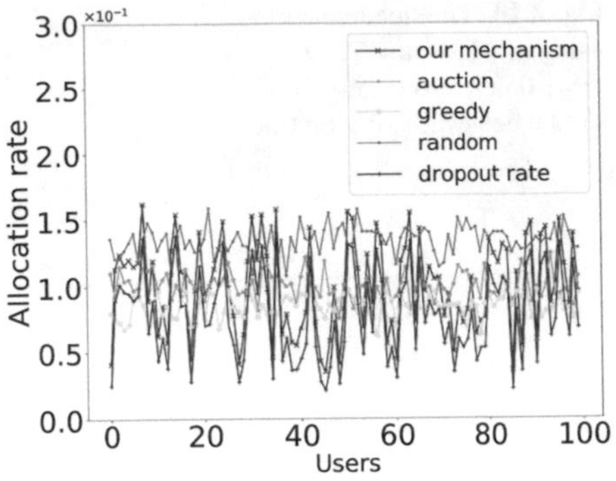

Fig. 2.12 The allocation rate performance comparison with baselines on the trace

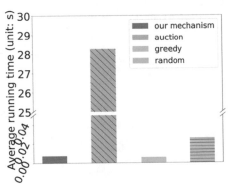

Fig. 2.13 The running time performance comparison with baselines on the trace

2.6   Conclusion

In this chapter, we design an incentive mechanism for crowdsensing under time-varying scenarios. The incentive mechanism is modeled by a three-stage Stackelberg game. Moreover, we leverage Lyapunov optimization to handle users' long-term participation issues and ensure the platform's time-average profits. Additionally, we develop an FPTAS for users to calculate their interests of tasks based on their mobile devices' computing capabilities. Simulation results validate the efficiency and effectiveness of our mechanism. In the future, we consider the impact of privacy and extend our mechanism generally, like Federated Crowdsensing [43].

References

1. Ganti, R.K., Ye, F., Lei, H.: Mobile crowdsensing: current state and future challenges. IEEE Commun. Mag. **49**(11), 32–39 (2011)
2. Guo, B., Liu, Y., Wang, L., Li, V.O.K., Lam, J.C.K., Yu, Z.: Task allocation in spatial crowdsourcing: current state and future directions. IEEE Internet Things J. **5**(3), 1749–1764 (2018)

3. Waze: Waze. https://www.waze.com
4. Gigwalk: Gigwalk. http://gigwalk.com/
5. Zheng, Y., Liu, F., Hsieh, H.P.: U-air: when urban air quality inference meets big data. In: Proceedings of the 19th SIGKDD conference on Knowledge Discovery and Data Mining (KDD 2013) ACM SIGKDD, pp. 1436–1444 (2013)
6. Wang, L., Zhang, D., Pathak, A., Chen, C., Xiong, H., Yang, D., Wang, Y.: CCS-TA: quality-guaranteed online task allocation in compressive crowdsensing. In: ACM International Joint Conference on Pervasive and Ubiquitous Computing (UbiComp), pp. 683–694 (2015)
7. Guo, B., Chen, H., Yu, Z., Xie, X., Huangfu, S., Zhang, D.: Fliermeet: a mobile crowdsensing system for cross-space public information reposting, tagging, and sharing. IEEE Trans. Mobile Comput. 14(10), 2020–2033 (2015)
8. Zhang, X., Yang, Z., Sun, W., Liu, Y., Tang, S., Xing, K., Mao, X.: Incentives for mobile crowd sensing: a survey. IEEE Commun. Surv. Tuts. 18(1), 54–67 (2016)
9. Feng, Z., Zhu, Y., Zhang, Q., Ni, L.M., Vasilakos, A.V.: Trac: truthful auction for location-aware collaborative sensing in mobile crowdsourcing. In: IEEE Conference on Computer Communications (INFOCOM), pp. 1231–1239 (2014)
10. Yang, D., Xue, G., Fang, X., Tang, J.: Crowdsourcing to smartphones: incentive mechanism design for mobile phone sensing. In: ACM International Conference on Mobile Computing and Networking (MobiCom), pp. 173–184 (2012)
11. Yang, D., Xue, G., Fang, X., Tang, J.: Incentive mechanisms for crowdsensing: crowdsourcing with smartphones. IEEE/ACM Trans. Netw. 24(3), 1732–1744 (2016)
12. Wen, Y., Shi, J., Zhang, Q., Tian, X., Huang, Z., Yu, H., Cheng, Y., Shen, X.: Quality-driven auction-based incentive mechanism for mobile crowd sensing. IEEE Trans. Veh. Technol. 64(9), 4203–4214 (2015)
13. Wei, Y., Zhu, Y., Zhu, H., Zhang, Q., Xue, G.: Truthful online double auctions for dynamic mobile crowdsourcing. In: IEEE Conference on Computer Communications (INFOCOM), pp. 2074–2082 (2015)
14. Jin, H., Su, L., Nahrstedt, K.: Centurion: incentivizing multi-requester mobile crowd sensing. In: IEEE Conference on Computer Communications (INFOCOM), pp. 1–9 (2017)
15. Karaliopoulos, M., Koutsopoulos, I., Spiliopoulos, L.: Optimal user choice engineering in mobile crowdsensing with bounded rational users. In: IEEE Conference on Computer Communications (INFOCOM), pp. 1054–1062 (2019)
16. Duan, Z., Li, W., Cai, Z.: Mutual-preference driven truthful auction mechanism in mobile crowdsensing. In: IEEE International Conference on Distributed Computing Systems (ICDCS), pp. 1233–1242 (2019)
17. Wang, L., Yu, Z., Zhang, D., Guo, B., Liu, C.H.: Heterogeneous multi-task assignment in mobile crowdsensing using spatiotemporal correlation. IEEE Trans. Mobile Comput. 18(1), 84–97 (2018)
18. Li, H., Li, T., Wang, W., Wang, Y.: Dynamic participant selection for large-scale mobile crowd sensing. IEEE Trans. Mobile Comput. 18(12), 2842–2855 (2018)
19. Nie, J., Xiong, Z., Niyato, D., Wang, P., Luo, J.: A socially-aware incentive mechanism for mobile crowdsensing service market. In: IEEE Global Communications Conference (GLOBECOM), pp. 1–7 (2018)
20. Cheung, M.H., Hou, F., Huang, J.: Make a difference: diversity-driven social mobile crowdsensing. In: IEEE Conference on Computer Communications (INFOCOM), pp. 1–9 (2017)
21. Xiao, L., Li, Y., Han, G., Dai, H., Poor, H.V.: A secure mobile crowdsensing game with deep reinforcement learning. IEEE Trans. Inf. Forens. Secur. 13(1), 35–47 (2018)
22. Zhang, X., Xue, G., Yu, R., Yang, D., Tang, J.: Truthful incentive mechanisms for crowdsourcing. In: IEEE Conference on Computer Communications (INFOCOM), pp. 2830–2838 (2015)
23. Chen, Y., Li, B., Zhang, Q.: Incentivizing crowdsourcing systems with network effects. In: IEEE Conference on Computer Communications (INFOCOM), pp. 1–9. IEEE, Piscataway (2016)
24. Zhan, Y., Xia, Y., Zhang, J.: Incentive mechanism in platform-centric mobile crowdsensing: a one-to-many bargaining approach. Comput. Netw. 132, 40–52 (2018)

25. Zhang, Y., Gu, Y., Pan, M., Tran, N.H., Dawy, Z., Han, Z.: Multi-dimensional incentive mechanism in mobile crowdsourcing with moral hazard. IEEE Trans. Mobile Comput. **17**(3), 604–616 (2018)
26. Jin, H., Guo, H., Su, L., Nahrstedt, K., Wang, X.: Dynamic task pricing in multi-requester mobile crowd sensing with markov correlated equilibrium. In: IEEE Conference on Computer Communications (INFOCOM), pp. 1063–1071 (2019)
27. DiPalantino, D., Vojnovic, M.: Crowdsourcing and all-pay auctions. In: Proceedings of the 10th ACM Conference on Electronic Commerce, pp. 119–128 (2009)
28. Chong, E.K.P., Zak, S.H.: An Introduction to Optimization, vol. 76. Wiley, Hoboke (2013)
29. Gao, L., Hou, F., Huang, J.: Providing long-term participation incentive in participatory sensing. In: IEEE Conference on Computer Communications (INFOCOM), pp. 2803–2811 (2015)
30. Li, F., Liu, J., Ji, B.: Combinatorial sleeping bandits with fairness constraints. In: IEEE Conference on Computer Communications (INFOCOM), pp. 1702–1710 (2019)
31. Yu, H., Cheung, M.H., Gao, L., Huang, J.: Economics of public Wi-Fi monetization and advertising. In: IEEE Conference on Computer Communications (INFOCOM), pp. 1–9 (2016)
32. Yu, H., Iosifidisy, S., Biying, L., Huang, J.: Market your venue with mobile applications: collaboration of online and offline businesses. In: IEEE Conference on Computer Communications (INFOCOM), pp. 1934–1942 (2018)
33. Sun, L., Pang, H., Gao, L.: Joint sponsor scheduling in cellular and edge caching networks for mobile video delivery. IEEE Trans. Multimedia **20**(12), 3414–3427 (2018)
34. Marjanović, M., Antonić, A., Žarko, I.P.: Edge computing architecture for mobile crowdsensing. IEEE Access **6**, 10662–10674 (2018)
35. Li, T., Qiu, Z., Cao, L., Li, H., Guo, Z., Li, F., Shi, X., Wang, Y.: Participant grouping for privacy preservation in mobile crowdsensing over hierarchical edge clouds. In: IEEE International Performance Computing and Communications Conference (IPCCC), pp. 1–8 (2018)
36. Vazirani, V.V.: Approximation Algorithms. Springer, Berlin (2013)
37. Neely, M.J.: Stochastic network optimization with application to communication and queueing systems. Synthesis Lect. Commun. Netw. **3**(1), 1–211 (2010)
38. Zinkevich, M.: Online convex programming and generalized infinitesimal gradient ascent. In: International Conference on Machine Learning (ICML), pp. 928–936 (2003)
39. Yu, H., Neely, M., Wei, X.: Online convex optimization with stochastic constraints. In: Advances in Neural Information Processing Systems (NeurIPS), pp. 1428–1438 (2017)
40. Neely, M.J., Yu, H.: Online convex optimization with time-varying constraints (2017). arXiv:1702.04783
41. Yu, H., Neely, M.J.: Learning aided optimization for energy harvesting devices with outdated state information. In: IEEE Conference on Computer Communications (INFOCOM), pp. 1853–1861 (2018)
42. Li, Y., Li, F., Yang, S., Zhou, P., Zhu, L., Wang, Y.: Supplementary: three-stage stack-elberg long-term incentive mechanism and monetization for mobile crowdsensing: an online learning approach (2020). https://www.dropbox.com/s/qwwi6vqh38ub3cg/manuscript_TNSE_supp.pdf?dl=0
43. Li, Y., Li, F., Zhu, L., Sharif, K., Chen, H.: A two-tiered incentive mechanism design for federated crowd sensing. CCF Trans. Pervasive Comput. Interact. **4**(4), 339–356 (2022)

Chapter 3
Fair Incentive Mechanism for Mobile Crowdsensing

Abstract In this chapter, we jointly address practical issues in the incentive mechanism for MCS to fairly incentivize high-quality users' participation, like (1) the platform has no knowledge about users' sensing qualities beforehand due to their private information. (2) The platform needs users' continuous participation in the long run, which results in fairness requirements. (3) It is also crucial to protect users' privacy due to the potential privacy leakage concerns (e.g., sensing qualities) after completing tasks. Particularly, we propose the three-stage Stackelberg-based incentive mechanism for the platform to recruit participants. In detail, we leverage combinatorial volatile multi-armed bandits (CVMAB) to elicit unknown users' sensing qualities. We use the drift-plus-penalty (DPP) technique in Lyapunov optimization to handle the fairness requirements. We blur the quality feedback with tunable Laplacian noise such that the incentive mechanism protects locally differential privacy (LDP). Finally, we carry out experiments to evaluate our incentive mechanism. The numerical results show that our incentive mechanism achieves *sublinear* regret performance to learn unknown quality with fairness and privacy guarantee.

Keywords Fair guarantee · Unknown quality · Stackelberg game · Combinatorial volatile multi-armed bandits · Locally differential privacy

3.1 Introduction

3.1.1 Motivations

Nowadays, the pervasive smart devices in our daily life have changed the way to collect sensing data, thus giving rise to a new sensing paradigm in the Internet of Things (IoT), called *mobile crowdsensing (MCS)* [1, 2]. Benefiting from the diverse sensors embedded in the hand-held phones and the high mobility of users, MCS enables many applications relying on large-scale sensing data, such as real-time traffic monitoring [3], indoor mapping [4], air quality monitoring [5], and influence maximization in social networks [6].

© The Author(s), under exclusive license to Springer Nature Singapore Pte Ltd. 2024
Y. Li et al., *Incentive Mechanism for Mobile Crowdsensing*, SpringerBriefs
in Computer Science, https://doi.org/10.1007/978-981-99-6921-0_3

Generally speaking, an MCS system usually consists of requesters, a platform, and users [1]. The platform resides in the cloud for serving the task requests from requesters. Once receiving the sensing tasks, then the platform recruits active users (i.e., participants) to perform the sensing tasks by providing rewards as incentives. Finally, the platform returns the sensing data collected from the participants to the requesters and obtains the corresponding payments from the requesters. To build an effective MCS system, the research efforts are devoted to addressing the concerns including participant recruitment [5, 7–9], incentive mechanism [10, 11]. However, the platform has the control of the sensing data in the MCS, which may result in a severe security issue (e.g., privacy leakage) especially when the platform is malicious. With the increasing awareness of data privacy and security, different countries adopt laws to strengthen data security protection, e.g., General Data Protection Regulation (GDPR) and China's Cyber Security Law. To still exploit the advantages brought by MCS under data privacy and security requirement, Privacy-Preserving Mobile Crowdsensing (PPMCS) [12–15] is increasingly advocated.

In this chapter, we focus on the PPMCS scenario where a trusted server is introduced. We show the PPMCS architecture in Fig. 3.1. The main interactions in PPMCS are similar to the typical MCS with a significant difference: the sensing data collected by the participants are maintained by the trusted server. In step 5, the participants send their collected data to the trusted server. And the trusted server can perform data curation to protect the privacy of the participants. In detail, the trusted server only generates and provides noisy feedback of participants' quality to the platform in step 6. Note that the trusted server can derive participants' quality

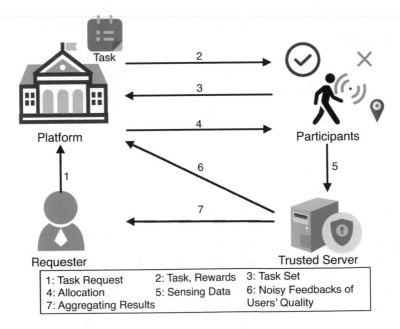

Fig. 3.1 Illustration of the focused Crowdsensing scenario

by evaluating their collected data. To better serve the requesters with high-quality sensing data, the trusted server also aggregates the sensing data (e.g., data cleaning, statistics) in step 7.

3.1.2 Challenges

To ensure the reliability of the PPMCS system, it is crucial for the platform to incentivize users with higher sensing qualities to perform the tasks, according to the designed incentive mechanism. However, one of the challenges here is that the users' sensing qualities are usually unknown to the platform due to privacy concerns.

One of the solutions to overcome the challenge of unknown quality information is to request the assistance of a trusted server. Since the trusted server maintains the collected sensing data of the participants, the trusted server can evaluate the quality of the sensing data and further infer users' true sensing qualities. However, due to privacy protection regulations, the trusted server has to generate random qualities in expectation of the true sensing qualities and send the noisy feedback to the platform. Although the noisy feedback of users' sensing qualities is used to help the platform to estimate users' unknown sensing qualities while circumventing privacy concerns, there are existing works indicating that users' true sensing qualities can be learned using online learning techniques (e.g., multi-armed bandits, MAB for short) even when obtaining the noisy feedbacks [7, 9, 16]. Thus, it is important to protect users' privacy and prevent users' true sensing qualities from being leaked. To the best of our knowledge, protecting users' true sensing qualities under noisy feedback is less investigated in the existing MCS literature. Although the existing works for crowdsensing focus on privacy preservation [12, 15, 17–24], they mainly work in the typical MCS architecture. In this chapter, we focus on PPMCS architecture with a trusted server where privacy protection is more controllable.

Except for the privacy factor, fairness is another important factor that should be incorporated into the incentive mechanism. When performing sensing tasks, all users desire to be recruited by the platform such that they have equal chances to earn the rewards. This poses a fairness requirement to the platform when designing a participant recruitment algorithm (PRA) in the incentive mechanism. Without ensuring fairness, some users with less recruiting may leave the crowdsensing system, which suppresses users' long-term participation.

Note that the existing works [7, 9, 10, 13, 16, 25, 26] only consider one or two challenges above but do not address unknown quality information, fairness requirement and privacy concerns together. We aim to jointly bridge this gap by fixing the problems arising in our PPMCS setting.

In Fig. 3.2, we present the challenges in our focused PPMCS scenario in this chapter, where different tasks are posted to the platform with corresponding payments at different rounds. In a round, the different users are active and choose one or more tasks as their interest set that are reported to the platform. Then, the platform invokes a PRA to recruit a subset of users to perform all the tasks.

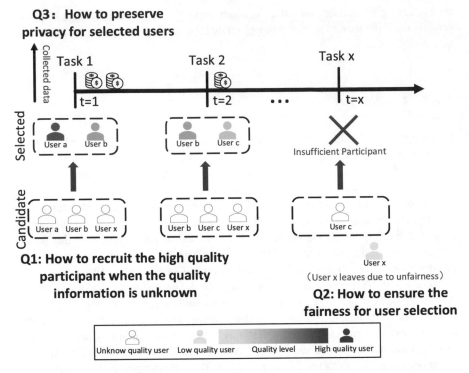

Fig. 3.2 The challenges in our scenario

The interactions involve problems that include, (1) How to recruit high-quality participants when the quality information is unknown. In the previous Stackelberg game framework for crowdsensing, the users' qualities are usually assumed to be known and incorporated into the utility function. However, the users' qualities are unknown in practice due to privacy concerns and requirements. How to estimate the users' unknown qualities is an important issue to enable the Stackelberg-based incentive mechanism for crowdsensing; (2) How to ensure fairness in user selections. Selecting high-quality users and ensuring fairness between all users will encounter a tradeoff problem between quality requirements and fairness guarantees. This issue becomes more challenging when the users' qualities are unknown, because the tradeoff problem becomes a regret-fairness tradeoff. How to balance this tradeoff involves the efficiency of learning the unknown qualities and guaranteeing fairness; (3) How to preserve privacy for the selected users. Accurately learning the users' unknown qualities and protecting users' private qualities will introduce a tradeoff problem between unknown quality issues and privacy concerns (i.e., regret-privacy tradeoff). How to balance this tradeoff also further affects the efficiency of learning the unknown qualities. We will model the incentive mechanism for PPMCS as a three-stage Stackelberg game [27] and derive the Stackelberg Equilibrium (SE) as an incentive mechanism design solution.

3.1.3 Contributions

Handling the unknown quality with noisy feedback involves the exploration-exploitation tradeoff (recruiting optimal users with the learned qualities in the past or recruiting potentially better users with fewer qualities information to reduce the uncertainty of the unknown quality), we incorporate the combinatorial volatile multi-armed bandits (CVMAB) technique [16, 28, 29] into the incentive mechanism to address the unknown quality issue with the objective of minimizing the regret brought by users' unknown sensing qualities. The reason why using CVMAB is that the platform recruits a subset of users while the ground user set is different (volatile) over time. Specifically, we consider the submodular function for tasks' completion due to the diminishing-return property of users' contributions on the completion qualities of tasks.

When taking fairness into account, we also have to address the regret-fairness tradeoff (minimizing the regret during learning the unknown qualities or ensuring fairness for less recruited users) [25]. When preserving privacy for users, we also have to resolve the regret-privacy tradeoff that involves the conflicting objectives between regret minimization for learning the unknown quality or preserving privacy. In this chapter, we aim to study CVMAB with submodular function while guaranteeing two conflicting objectives (i.e., regret-fairness and regret-privacy) and achieving two theoretical tradeoffs regarding the sublinear regret bound. Then we apply the studied theoretical framework to the focused PPMCS scenario. Our contributions include,

- We propose a three-stage Stackelberg incentive mechanism for the strategic interactions in our PPMCS scenario. In our incentive mechanism, we jointly consider quality, fairness, and privacy concerns.
- Specifically, we propose a PRA that jointly addresses unknown quality information, fairness requirement and privacy concerns together and makes theoretical tradeoffs among three objectives (regret, fairness and privacy). Our PRA achieves regret bound $O(\alpha\sqrt{bNT\log T})$, time-average regret bound $O(\alpha N\frac{\log T}{T})$ and ς-locally differential privacy (LDP) regret bound $O(\alpha\varsigma^{-1}\sqrt{bNT\log T})$, which is sublinear in terms of time span T, the number of users N, the maximum recruiting number b. $\varsigma > 0$ is the privacy budget. α is the approximation ratio to the submodular maximization in the user selection problem.
- We conduct simulations to evaluate the proposed algorithms to verify the theoretical results over different benchmarks.

3.1.4 Related Work

As a new sensing paradigm, most works [2, 5, 30–33] pay attention to study the task allocation problem in MCS. When studying task allocation with reward (i.e., pricing), the problem becomes participant recruitment [8, 9, 26, 34, 35]. And the improved version of participant recruitment is referred to as the incentive mechanism. In this chapter, we focus on incentive mechanism design for MCS.

The incentive mechanisms for MCS include Stackelberg-based incentive mechanism [36–38], Auction-based incentive mechanism [8, 10] and Other games based incentive mechanism (like network effect [39]), contracting [40]), and Markov game [41]). When different distributed and mutual-preference auction-based incentive mechanisms [42–44] are proposed, the users' quality, fairness, and privacy concerns are not considered in these work, which is jointly studied in our work. In this chapter, we focus on the three-stage Stackelberg incentive mechanism where participant recruitment and reward determination are considered in different stages. When the works [45, 46] also focus on the Stackelberg-based incentive mechanism. The differences between the works [45, 46] and our work are discussed as follows. (1) The suggested works [45, 46] formulate the strategic interactions between the platform and users as Markov Decision Process (MDP) and model the dynamic impacts (e.g., users' uncertain resource demands, dynamic availability) into the MDP. Then, they use the Deep Reinforcement Learning (DRL) technique to solve the Stackelberg Equilibrium and obtain the participation strategy and pricing strategy. However, they do not consider the unknown quality, fairness issues and privacy issues, which are jointly considered in our work. The existing works [7, 9] also use the MAB technique to study the participant recruitment problem. However, they do not consider submodular reward, and fairness constraints. Gao et al. [10] propose a fairness-aware incentive mechanism to guarantee users' long-term participation. However, privacy protection is negligible in their incentive mechanism, which may continuously lead to users' privacy leakage due to long-term participation.

Knowing the importance of privacy protection in MCS, many works [12–15] propose different privacy protection schemes in different MCS scenarios. Lin et al. [12] design two score functions that have differential privacy properties and then propose an auction-based incentive mechanism. However, our incentive mechanism is designed based on the Stackelberg game, which is quite different from auction essentially. In addition, we consider the impact of the unknown qualities of users in incentive mechanism design. Li et al. [13] propose a privacy-preserving grouping method to protect users' privacy in participant selection. Zhang et al. [14] study the privacy-preserving truth discovery for MCS. However, we study the privacy-preserving incentive mechanism for MCS in this chapter. The existing work [15] also studies the Stackelberg-based incentive mechanism in privacy-preserving MCS and investigates the impact of the uncertainty on their incentive mechanism based on reinforcement learning (RL). However, they do not take fairness constraints into account. In this chapter, we aim to jointly address unknown quality information, fairness requirement and privacy concerns together for constructing reliable PPMCS.

3.2   Game Modeling

We first describe our studied setting of PPMCS, then introduce the parameters and strategies of the user and the platform. Finally, we state the studied problems. The mathematical notations are summarized in Table 3.1.

3.2.1   Overview

We consider a time-evolving crowdsensing scenario with a time span of T. We call unit time as time slot and round interchangeably. During each time slot, the platform interacts with the requesters and users subsequently, where their strategic interactions are modeled using the three-stage Stackelberg game in this chapter (to be detailed later). We consider M requesters in our setting (i.e., $M = \{1, 2, \ldots, M\}$). At time slot t, requester j has a task demand and sends it to the

Table 3.1 Notations used in Chap. 3

Notations		Meanings
Users	$\mathcal{N}, N$	User set and its size
	q_i	The sensing quality of user i in expectation
	c_{ij}	Sensing cost for user i perform task j
	Γ_i^t	Maximum number of tasks user i can perform[a]
	S_i^t	The interest set of user i
	$Z_i(t)$	The backlog of virtual credit queue for user[a]
	$\mathbf{Z}(t)$	The virtual credit queue vector in round t
Platform	r_j^t	The reward variable platform prices for task[a]
	b_j^t	At most b_j^t users platform selects for task[a]
	$\mathcal{N}_j^t$	The user set contains users interested in task[a]
	d_i	The cardinality constraint platform imposes on S_i^t
	$\mathcal{P}_j^t$	User set platform determines to perform task[a]
	$x_{i,j}^t$	Variable implies whether user i performs task[a]
	V	Fairness control parameter
Tasks	$\mathcal{M}, M$	Requester set and its size
	$\mathcal{M}_t$	The task set in round t
	k_j^t	The cover number of task j in round t
	p_j^t	The payment requester j makes in round t
	γ_j^t	The hardness parameter of task j
	$f_j(\cdot)$	The complete quality function of task j
	D_j^t	The delay task j can tolerates in round t

[a] Without confusion, we omit the expression of "in round t"

platform. Then, the platform recruits users to perform the task and collects the sensing data from them. In this chapter, we assume that each requester can only demand a task at each time slot. As some requesters may not have task demands at time slot t, we denote by $\mathcal{M}_t \subseteq \mathcal{M}$ the task set arrived in round t. Complementing to request for task j, requester j also declares a payment p_j^t to the platform. Besides, the platform and requester j have consensus to characterize the hardness level $\frac{1}{\gamma_j^t}$ of task j, where $\gamma_j^t > 0$ is the hardness parameter. The higher γ_j^t is, the less difficult task j is completed. For example, collecting multi-media data is harder than sensing inertial data since more operational costs are incurred in the former one. Besides, γ_j^t is related to the privacy-preserving level imposed on the incentive mechanism (e.g., ς-local differential privacy). By paying p_j^t, requester j has quality requirements. To ensure the quality requirement of task j, the platform has to recruit at least k_j^t users to perform task j. After assigning all tasks in $\mathcal{M}_t$, the strategic interactions in round t are finished. Note that the participants have to take the time to perform the assigned tasks. To characterize the real-time requirement of tasks, we introduce tasks' delay, after which the participants should complete the tasks and return the corresponding sensing data. We denote by D_j^t the delay of task j. The smaller D_j^t is, the more urgent task j is.

3.2.2   User Model

We consider user set $\mathcal{N} = \{1, 2, \ldots, N\}$. A quality attribute $q_i \in [0, 1]$ is associated with each user $i \in \mathcal{N}$. The larger q_i suggests that user i performs a task with higher sensing quality. However, q_i is private information of user i, which is not known a priori for the platform. This leads to a challenge to design the optimal participant recruitment strategy for the platform. Fortunately, the feedback of the obtained sensing data from the trusted server can be seen as the noise feedback of q_i. The participant recruitment algorithm can utilize this feedback to learn the users' quality information like recommendation systems in an online learning manner.

Limited by the remaining battery and sensing resources, user i can at most perform Γ_i^t tasks in round t. Due to the hardness (i.e., reflected by γ_j^t) and the allowed delay D_j^t of task j, the sensing cost for user i performing task j is c_{ij}. When the platform sends the current tasks $\mathcal{M}_t$ together with the priced rewards to the active users, each user $i \in \mathcal{N}$ selects a subset of tasks $\mathcal{S}_i^t$ ($|\mathcal{S}_i^t| \leq \Gamma_i^t$) as *interest set* based on her limited resources. Then, user i reports her interest set $\mathcal{S}_i^t$ back to the platform for further recruiting decisions. Due to the selfishness and rationality of users, they tend to select a task with a higher reward and less sensing cost into the interest set.

3.2.3 Platform Model

To provide sensing services, the platform has to complete two procedures for the strategic interactions: (1) pricing the rewards, (2) and selecting the users.

Reward Determination For each task $j \in \mathcal{M}_t$, the platform has to determine a proper reward r_j^t to incentivize users to perform this task. The platform's goal is to make some profits when pricing the rewards. It implies that the total rewards should be less than the payment p_j^t when the participants complete task j. In a practical scenario, the number of recruited participants for task j is no more than b_j^t but larger than the covering number k_j^t (i.e., $b_j^t \geq k_j^t$). Thus, the reward pricing is associated with the following budget constraint,

$$b_j^t r_j^t \leq p_j^t. \tag{3.1}$$

For any user $i \in \mathcal{N}$, the following factors determine whether user i adds task j into her interest set (i.e., tasks j is covered by user i): sensing cost c_{ij}, sensing resource restriction (i.e., Γ_i^t), and the hardness of task j (i.e., γ_j^t). However, these factors are heterogenous over different users and tasks. We can define a probability of the event that task j is covered using exponential style probability model,

$$\mathbb{P}[\text{task } j \text{ is covered}] = 1 - e^{-\gamma_j^t r_j^t}. \tag{3.2}$$

This kind of probability is commonly used in mobile data monetization [47], which presents a similarity to reward pricing in the MCS. From Eq. (3.2), we can see that the larger γ_j^t and r_j^t are, the more likely task j is covered and vice versa. Except for the budget constraint, the platform has to ensure the participation constraint, since the requester j requires at least k_j^t users to participate in task j for quality guarantee. The participation constraint can be expressed as,

$$\mathbb{P}[\text{task } j \text{ is covered}] \geq \frac{k_j^t}{N}. \tag{3.3}$$

Equation (3.3) indicates that the task j is covered by at least k_j^t users in expectation by pricing reward r_j^t.

Participant Selection After collecting the interest sets from users, the platform can determine a user set only containing the users who are interested to participate in task j, called covering set $\mathcal{N}_j^t \subseteq \mathcal{N}$. The platform recruits a set of users $\mathcal{P}_j^t \subseteq \mathcal{N}_j^t$ ($|\mathcal{P}_j^t| \leq b_j^t$) to perform task j such its completion quality is maximized. We define a task's completion quality function as a set function $f_j : 2^{\mathcal{N}_j^t} \rightarrow \mathbb{R}$, where its value is dependent on the users' sensing qualities regarding a specific user set. In particular, we consider that $f_j(\cdot)$ has a diminishing return property. This means that the more users the platform recruits, the less the marginal return of the task's completion quality becomes, which commonly appears in crowdsensing and crowdsourcing areas [11, 16]. The diminishing return property implies $f_j(\cdot)$ is a submodular function. Given any user set A and B ($A \subseteq B \subseteq \mathcal{N}^t_j$), the following inequality holds for the submodular function $f_j(\cdot)$,

$$f_j(A \cup \{i\}) - f_j(A) \geq f_j(B \cup \{i\}) - f_j(B), \tag{3.4}$$

for any user $i \in \mathcal{N}^t_j \backslash B$. For ease of exposition, we define a boolean variable $x^t_{i,j} \in \{0, 1\}$ to indicate whether the platform recruits user i to perform task j. If user $i \in \mathcal{P}^t_j$, we have $x^t_{i,j} = 1$, otherwise $x^t_{i,j} = 0$.

3.2.4   Fairness Model

Considering multiple strategic interactions within a time horizon T, if a user is not recruited as a participant in the long term, she will lose the chance to earn the rewards. Thus, the long-term unselected users will drop out of the MCS system. To ensure all users' long-term participation, it is necessary to propose a participant recruitment algorithm with a fairness guarantee. We define the fairness model as follows,

Definition 3.1 A participant recruitment algorithm $\mathcal{A}$ is said to be fair if and only if the following inequality holds for each user i.

$$\liminf_{t \to \infty} \frac{\sum^t_{\tau=0} \sum_{j \in \mathcal{M}_t} x^t_{i,j}}{t} = \overline{x}_i \geq F_i, \forall i \in \mathcal{N}. \tag{3.5}$$

Definition 3.1 indicates that the time-average participation rate $\overline{x}_i$ is no less than a given threshold F_i for each user $i \in \mathcal{N}$, when algorithm $\mathcal{A}$ has fairness guarantee. Similar definitions can be found in [10, 25]. In Eq. (3.5), we can see that the fairness guarantee relies heavily on the threshold F_i. One way to determine threshold F_i is to associate it with the learned quality information about q_i. It ensures that high-quality users are recruited more than low-quality users. However, the fairness guarantee is conflicting to maximize the task's completion quality, since low-quality users may be recruited to guarantee fairness in some rounds. To balance this tradeoff, we utilize Lyapunov optimization [48] to design the participant recruitment algorithm and analyze the guaranteed performance later.

Note that the fairness studied in our incentive mechanism means scheduling fairness. With the fairness guarantee, our incentive mechanism can ensure that all users have opportunities to participate in performing tasks and earn rewards. This is important for the incentive mechanism to induce users' long-term participation. In this work, the payment has an implicit connection with the users' sensing qualities. According to our fairness definition, high-quality users are selected more times in our incentive mechanism. Therefore, they will receive more payments.

3.2.5 Privacy-Preserving Model

We aim to design our participant recruitment algorithm for MCS with a ς-local differential privacy (ς-LDP) guarantee. We use the LDP definition in [49, 50] in MCS as follows.

Definition 3.2 Considering task j, a participant recruitment algorithm $\mathcal{A}$ is said to be ς-local differential private or ς-LDP, if for any user quality information sets Q_1, Q_2 that differ in at most one entry, and user set $\mathcal{P} \subset 2^{\mathcal{N}_j}$, we have,

$$\mathbb{P}[\mathcal{A}(Q_1) \in \mathcal{P}] \leq e^{\varsigma} \mathbb{P}[\mathcal{A}(Q_2) \in \mathcal{P}], \tag{3.6}$$

for given privacy parameter ς.

When $\varsigma = 0$, algorithm $\mathcal{A}$ provides the strongest privacy guarantee. However, the learning performance will become worst in this case of $\varsigma = 0$. When $\varsigma = \infty$, algorithm $\mathcal{A}$ has no privacy guarantee. ς-LDP enables perturbing users' noisy feedback (i.e., quality scores) such that the users' private sensing qualities information cannot be easily inferred and revealed. To protect ς-LDP, Laplacian and Gaussian mechanisms are usually used to perturb participants' quality scores with random noises by the parameter $1/\varsigma$.

3.2.6 Problem Statement

In this chapter, we study the incentive mechanism within T rounds' interactions problem, which is modeled as a three-stage Stackelberg game where the game players are the platform and users. The game players' tree-stage strategic interactions include: *Stage I: Platform's tasks rewards pricing problem, Stage II: Users' interest set determining problem*, and *Stage III: Platform's participant recruiting problem*. As mentioned above, the three subproblems are coupled together. We use backward induction [27, 51] to analyze the formulated Stackelberg Equilibrium (SE) strategies from stage III to stage I. In the next three sections, we first provide the problem formulation and present the potential challenges prior to proposing the corresponding algorithms.

3.3 Detailed Design

3.3.1 Stage III: Platform's Participant Recruitment Strategies

In stage III, the platform solves the participant recruiting problem, which can be formulated as submodular maximization. Note that when the platform's optimization

goal in this stage should maximize the total (weighted) completion qualities of all
requesters in our scenario, the objective can be equivalently transformed into the one
that maximizes one requester's achieved qualities. This is because the optimization
problem is decoupled with tasks and rounds. Different active users are different in
each round and the covering set $\mathcal{N}_j^t$ is also different for different tasks and different
rounds. Taking task $j \in \mathcal{M}_t$ as an example, the formulation is,

$$\max \quad f_j(\mathcal{P}_j^t)$$

$$(S3PS) \qquad \begin{cases} |\mathcal{P}_j^t| \leq b_j^t, \\ \overline{x}_i \geq F_i, \\ (\mathcal{P}_j^t \subseteq \mathcal{N}_j^t, i \in \mathcal{N}_j^t), \end{cases}$$

where the first inequality is cardinality constraint and the second one is fairness
constraint.

Lemma 3.1 *Problem (S3PS) is NP-hard.*

Proof Submodular maximization with cardinality constraint is NP-hard problem
[52, 53]. And our formulation has additional fairness constraints. Thus, it is also
NP-hard.   □

Due to the NP-hardness of problem (S3PS), it is indispensable to propose an
approximation algorithm. We will adapt the conventional random greedy algorithm
[53] to our problem while carefully handling the fairness constraints. In summary,
the challenges to solve problem (S3PS) are two-fold: (1) Users' quality information
is unknown to the platform, which leads to the issue that the objective has no closed
form; (2) Essentially, the fairness constraints may be conflicting with the objective
in some rounds, because ensuring fairness requires selecting the low-quality users.

Fortunately, in our PPMCS architecture, the trusted server can generate noisy
feedback to the platform by-product when aggregating the sensing results. Thus,
we address the first challenge by means of multi-armed bandits (MAB) to learn
users' sensing qualities information with noisy feedback in an online learning
manner. Specifically, we use combinatorial volatile multi-armed bandits (CVMAB)
with the "submodular reward function" to reformulate the participant recruitment
problem. For the second challenge, we leverage Lyapunov optimization to handle
the time-average fairness constraints. This requires subtly integrating drift-plus-
penalty (DPP) in Lyapunov optimization with UCB policy in MAB.

3.3.1.1   UCB-Based Participant Recruitment Algorithm

As mentioned before, we use CVMAB to reformulate the participant recruitment
problem due to the unknown sensing qualities of users. Taking task $j \in \mathcal{M}_t$ as an
example, the platform recruits the participants $\mathcal{P}_j^t$ from $\mathcal{N}_j^t$, which can be interpreted
as pulling arms $\mathcal{P}_j^t$. When the participants complete the assigned tasks and return

the sensing data to the trusted server, the platform can obtain the quality scores (e.g., the noisy level of acoustic data, the blurry level of the sensed pictures) regarding the users, which can be seen as the bandit feedbacks from the trusted server. When the platform recruits users $\mathcal{P}_j^t$ to perform task j, the quality scores about users (denoted by β_{ij}^t) are not only revealed by the platform, but also the overall score (denoted by β_{Pj}^t) about the objective $f_j(\mathcal{P}_j^t)$ is determined. This kind of bandit feedback is called semi-bandit feedback [28, 29]. The considered semi-bandit feedback enables us to utilize arm dependency among different user sets recruited by the platform. For example, we consider that P_1, P_2 are recruited at different time slots. When participants (P_1 or P_2) are returned the sensing data, the bandit feedbacks of $P_1 \cap P_2$ can be made full use of. We make an assumption about the quality scores as follows.

(Expected quality score) For user $i \in N$ performing task j, its quality score β_{ij}^t is equal to q_i in expectation, i.e., $\mathbb{E}[\beta_{ij}^t] = q_i$. Moreover, the overall quality is equal to $f_j(\mathcal{P}_j^t)$ in expectation when participants $\mathcal{P}_j^t$ are recruited.

Assumption 3.3.1.1 points out that the quality score β_{ij}^t is sampled from the distribution with mean q_i in an i.i.d manner. it is common in Crowdsensing and Crowdsourcing [7, 9, 16, 25], as the more quality scores are sampled, the smaller the gap between the quality average and the mean becomes.

Although we model the problem using the existing CVMAB, the difference is that the bandit feedback is not obtained immediately when the platform recruits a subset of users. This is because the participants have to take several rounds to perform tasks and return the sensing data. Thus, our bandit feedbacks are delayed several rounds before they are obtained. This is referred to as delayed bandit feedback [54].

Like the existing bandit framework, we use *regret* as a metric to evaluate the performance of the proposed algorithm, which is the difference between the cumulative maximum objective value under optimal arm $\mathcal{P}_j^{*t}$ in hindsight and the cumulative objective value $\mathcal{P}_j^t$ returned by the algorithm. However, it is unfair to use $f_j(\mathcal{P}_j^{*t})$ to define the regret due to the NP-hardness of problem $\mathcal{P}_j^{*t} = \arg\max_S f_j(S)$. Therefore, we use α-regret $\text{Reg}_\alpha(T)$ instead, which is calculated within T rounds as,

$$\text{Reg}_\alpha(T) = \alpha T f_j(\mathcal{P}_j^{*t}) - \sum_{t=1}^{T} f_j(\mathcal{P}_j^t), \tag{3.7}$$

where α is the approximation ratio of the approximation algorithm of problem $\mathcal{P}_j^{*t} = \arg\max_S f_j(S)$ ($0 \le \alpha \le 1$). When $f_j(\cdot)$ is monotone, then the pure greedy algorithm provides $\alpha = 1 - \frac{1}{e}$ approximation [52]. When $f_j(\cdot)$ is non-monotone, then the random greedy algorithm provides $\alpha = 0.372$ approximation [53]. The goal of the proposed participant algorithm is to minimize the regret in Eq. (3.7) such that the *sublinear* regret bound is achieved. The sublinear regret bound implies that the recruiting decision in round T is asymptotically optimal when T is sufficiently large (i.e., $\lim_{T \to \infty} \frac{\text{Reg}_{\alpha}(T)}{T} = 0$). This is because the platform can learn the users' quality information after T rounds.

To achieve sublinear regret bound, the platform has to balance the exploration-exploitation tradeoff that is whether to recruit seemingly good users as participants or to recruit less explored users with probably higher sensing qualities. Following the principle of "optimism in the face of uncertainty" [55], we calculate an upper confidence bound (UCB) for each user as the optimistic estimate of her sensing quality. Then we propose a combinatorial UCB-based participant recruitment algorithm. Note that the UCB implicitly integrates exploration and exploitation into a value. Let n_i be the recruiting number of user i up to round t. The UCB value for user i is calculated as,

$$\bar{q}_i = \min\{\hat{q}_i + \sqrt{\frac{\eta \ln(t)}{n_i}}, 1\}, \tag{3.8}$$

where the first term $\hat{q}_i$ is the mean quality score of user i, while the second term $\sqrt{\frac{\eta \ln(t)}{n_i}}$ is confidence radius that increases with the round t (i.e., exploration) but decreases with the recruiting number (i.e., exploitation). Parameter η is the exploring rate. The mean quality score $\hat{q}_i$ of user i can be derived by using $\hat{q}_i = \sum_{\tau=1}^{t} \mathbb{I}[x_{i,j}^t = 1]\beta_{ij}^t / n_i$. After obtaining the UCB value $\bar{q}_i$ for each $i \in \mathcal{N}_j^t$, the platform can use the random greedy algorithm (RGA) as an offline oracle to output the recruiting decision for task j. The reason why using the RGA as an oracle is that RGA can handle monotone $f_j(\cdot)$ with $1 - \frac{1}{e}$ approximation and non-monotone $f_j(\cdot)$ with 0.372 approximation. By integrating RGA and delayed bandit feedback with the UCB policy, we can propose a UCB-based participant recruitment algorithm for task j in round t. The pseudo-code of the proposed algorithm is presented in Algorithm 3.1.

Algorithm 3.1 takes the covering set $\mathcal{N}_j^t$, the mean quality set $\{\hat{q}_i\}_i$ and the recruiting number set $\{n_i\}_i$ as input, and outputs the recruiting set $\mathcal{P}_j^t$ as a decision for task j in round t. In lines 1–3, UCB-PRA is to check whether to enter a cold start initialization, which randomly chooses a subset containing user i due to no exploited samples for calculating the mean (i.e., $n_i = 0$). In line 4–5, UCB-PRA calculates the UCB value $\bar{q}_i$ for each user $i \in \mathcal{N}_j^t$ according to Eq. (3.8). In line 6, UCB-PRA initializes $\mathcal{P}_j^t$ to be an empty set. In lines 7–16, UCB-PRA runs a random greedy policy to derive the sub-optimal recruiting set $\mathcal{P}_j^t$ with UCB values $\{\bar{q}_i\}_i$. In line 18, UCB-PRA updates the statistics regarding users (i.e., $\{\hat{q}_i\}_i$ and $\{n_i\}$), once receiving the sensing data from users (i.e., the delayed feedback). Finally, the recruiting set $\mathcal{P}_j^t$ is returned in line 19. We can see that the computation complexity of UCB-PRA is $O(b_j^t |\mathcal{N}_j^t|)$. The effectiveness of UCB-PRA is supported by the Chernoff-Hoeffding inequality [55] which is $\mathbb{P}[|\hat{q}_i - q_i| > \sqrt{\frac{\eta \ln(t)}{n_i}}] < 2 \exp(-2\eta \ln t)$. It implies that the more user i has been recruited, the higher probability ground-truth quality q_i lies in the confidence range related to $\hat{q}_i$. Note that the Chernoff-Hoeffding inequality

---

Algorithm 3.1 UCB-based participant recruitment algorithm (UCB-PRA) for task j in round t

---

Input: the covering set $\mathcal{N}_j^t$, the mean quality set $\{\hat{q}_i\}_i$ and the recruiting number set $\{n_i\}_i$

Output: the recruiting set $\mathcal{P}_j^t$

1: **if** $n_i = 0$, $\exists i \in \mathcal{N}_j^t$ **then**
2:      Randomly choose a subset containing user i: $\mathcal{P}_j^t \subseteq \mathcal{N}_j^t$, $|\mathcal{P}_j^t| \le b_j^t$
3: **else**
4:      Calculate $\bar{q}_i$, $\forall i \in \mathcal{N}_j^t$ according to Eq. (3.8)
5:      Based on $\{\bar{q}_i\}_i$, determine completion quality function $f_j(\cdot)$
6:      Let $\mathcal{P}_j^t \leftarrow \emptyset$
7:      **for** $i = 1$ **to** b_j^t **do**
8:          **if** $i \le \lceil 0.21 b_j^t \rceil$ **then**
9:              $k \leftarrow 2(b_j^t - i + 1)$
10:          **else**
11:              $k \leftarrow b_j^t$
12:          **end if**
13:          Greedy to choose a subset $M \subseteq \mathcal{N}_j^t \backslash \mathcal{P}_j^t$ to maximize $\sum_{i \in M} \Delta(i|\mathcal{P}_j^t)$, where $\Delta(i|\mathcal{P}_j^t) = f_j(\mathcal{P}_j^t \cup \{i\}) - f_j(\mathcal{P}_j^t)$, $|M| = k$
14:          Randomly select a user i from M
15:          $\mathcal{P}_j^t \leftarrow \mathcal{P}_j^t \cup \{i\}$
16:      **end for**
17: **end if**
18: Update $\hat{q}_i$ and n_i once receiving the sensing data from user $i \in \mathcal{N}$
19: **return** $\mathcal{P}_j^t$.

---

plays a role in proving the regret upper bound of UCB-PRA which is presented in the following theorem.

Theorem 3.1 *Considering T rounds' interactions in MCS, UCB-PRA incurs the α-regret to learn users' qualities information that is upper bound as follows,*

$$
\begin{aligned}
Reg_\alpha(T) \le & 93J\sqrt{bNT\ln(T)} + \alpha JN \ln(T) \\
& + (1 + \zeta(2\eta - 1))\alpha JN,
\end{aligned} \tag{3.9}
$$

where $\zeta(\cdot)$ is Riemann zeta function and $b = \max_{j,t} b_j^t$ and $J = \alpha \max_j \max_S f_j(S)$. When $\eta = 1.5$, $\zeta(2) = \frac{\pi^2}{6}$.

Proof The first term and third term follow the combinatorial MAB with Chernoff-Hoeffding bound [29]. The second term is due to the delayed feedback [54]. For the heterogeneity in different rounds, we scale the variables b_j^t, $f_j(\cdot)$, and $|\mathcal{N}_j^t|$ to b, J, and N. □

Remark 3.1 Theorem 3.1 demonstrates that UCB-PRA achieves sublinear regret performance $O(\alpha\sqrt{bNT\ln(T)})$. Thus, UCB-PRA is asymptotically optimal in terms of learning the unknown qualities of information. Compared to [9, 16] with $O(b)$ and $O(b^2)$, our UCB-PRA achieves a regret bound $O(\sqrt{b})$ in terms of the maximum recruiting number b. Although our UCB-PRA is order-optimal like [25, 29] with respect to time span T and the maximum recruiting number b, UCB-PRA can handle the delayed feedback.

Note that the UCB-PRA algorithm only considers the recruiting decision for a single task j. Independently performing UCB-PRA for each task $j \in \mathcal{M}_t$ is inefficient because the arm dependency among different tasks is not made full use of. Actually, a user i may perform different tasks (i.e., $\exists i \in \mathcal{N}_{j_1}^t, i \in \mathcal{N}_{j_2}^t, \mathcal{N}_{j_1}^t \cap \mathcal{N}_{j_2}^t \neq \emptyset$) and return more than 2 sensing data, which causes more bandit feedbacks. One way to address this issue is to create a shared memory that stores users' statistics (i.e., $\{\hat{q}_i\}_i$ and $\{n_i\}$). And we design a new algorithm (denoted by UCB-PRA-I) that invokes UCB-PRA with accessing the shared memory to handle all tasks' recruiting decisions. UCB-PRA-I can calculate all tasks' recruiting decisions in a parallel or sequential manner. The regret performance of UCB-PRA-I is presented as,

Theorem 3.2 Let $M' = \max_{j_1, j_2, t} \max\{|\mathcal{N}_{j_1}^t \cap \mathcal{N}_{j_2}^t|\}$. the regret of UCB-PRA-I is upper bounded as,

$$
\begin{aligned}
Reg_\alpha(T) \leq & 93J \frac{\sqrt{bNT \ln(T)}}{M'} + \alpha JN \ln(T) \\
& + \frac{(1 + \zeta(2\eta - 1))\alpha JN}{M'},
\end{aligned}
\tag{3.10}
$$

Proof The proof is similar to Theorem 3.1 by considering the arm dependency among different tasks. However, arm dependency does not influence the delayed feedback.   □

3.3.1.2   LyaUCB Based Participant Recruitment Algorithm

The second challenge of problem (S3PS) is the imposed fairness constraints in Eq. (3.5) that may be conflicting with the objective, i.e., $f_j(\cdot)$. Therefore, the UCB-based participant recruitment algorithm does not only balance the exploration-exploitation tradeoff, but also controls the regret-fairness tradeoff [25]. In particular, we consider that a general submodular objective function (e.g., monotone or non-monotone) while reference [25] only focuses on linear set function. It implies that the pure greedy algorithm cannot handle the non-monotone case. To this end, we adapt the algorithm with a random greedy algorithm by considering fairness constraints.

Since the fairness constraints are represented as the time-average inequalities related to the previous recruiting decisions in the past, it is necessary to decouple the fairness constraints at different rounds. Leveraging the Lyapunov optimization, we define a virtual credit queue for each user i whose backlog in round t is denoted as $Z_i(t)$ with initial value $Z_i(0) = 0$. Queue $Z_i(t)$ is updated according to the following inequality, which can be transformed by Eq. (3.5),

$$Z_i(t+1) = \max\{Z_i(t) - x_{i,j}^t + \theta F_i, 0\}, \tag{3.11}$$

where θ is a factor to scale the fairness threshold F_i. The $Z_i(t)$ value implicitly contains the statistical information related to past decisions. Eq. (3.11) can be interpreted as taking F_i into the queue and popping $x_{i,j}^t$ out of the queue. Let $\mathbf{Z}(t)$ be the queue vector. Then the fairness constraints are satisfied if and only if the virtual credit queue system $\mathbf{Z}(t)$ is stable. The stable queues $\mathbf{Z}(t)$ implies that the queues' backlogs are bouned (i.e., $\limsup_{t \to \infty} \sum_{\tau=0}^{t} \mathbb{E}[\sum_{i=1}^{N} Z_i(\tau)] < \infty$).

Like vector norm, we define a Lyapunov function [48] to reflect the "stability" of the queues $\mathbf{Z}(t)$ in round t as,

$$L(\mathbf{Z}(t)) = \frac{1}{2} \sum_{i=1}^{N} \omega_i Z_i^2(t), \tag{3.12}$$

where $\omega_i > 0$ is a weight of importance for user i. We consider that the platform fairly treats all users by setting $\omega_i = 1$, $\forall i \in N$. while directly stabilizing $\mathbf{Z}(t)$ is implicit, it is useful to achieve stability by minimizing the Lyapunov drift, which is defined as the expected difference of the Lyapunov function between round $t+1$ and t given the current queues $\mathbf{Z}(t)$. Lyapunov drift can be formally represented as $\Delta(\mathbf{Z}(t)) \overset{\Delta}{=} \mathbb{E}[L(\mathbf{Z}(t+1)) - L(\mathbf{Z}(t))|\mathbf{Z}(t)]$. According to drift-plus-penalty technique in Lyapunov optimization, to jointly minimize the regret of recruiting users while guaranteeing the fairness constraints, the objective becomes $\mathcal{P}_j^{*t} = \arg\max_S \Delta(\mathbf{Z}(t)) + V\text{Reg}_\alpha(t)$ where V is a tuneable parameter regarding to regret-fairness tradeoff.

Based on the above notations, we can characterize the participant recruitment policy with a fairness guarantee in the following lemma.

Lemma 3.2 *Given parameter $V > 0$, let $V\bar{q}_i + Z_i(t)$ be Lyapunov UCB value for user i and $f_j'(S) = f_j(S) + V\sum_{i \in S} Z_i(t)$ be the Lyapunov completion quality function for task j. Then the recruiting policy with fairness guarantee is $\mathcal{P}_j^t = \arg\max_S \alpha f_j(\cdot)$, where α is an approximation ratio.*

Proof This lemma can be draw by expanding the drift-plus-penalty function $\mathcal{P}_j^t = \arg\max_S \Delta(\mathbf{Z}(t)) + V\text{Reg}_\alpha(t)$. And we also consider the fact of $(\max\{Q - x + D, 0\})^2 \leq Q^2 + x^2 + D^2 + 2Q(D - x)$ to expand $\Delta(\mathbf{Z}(t))$ and UCB policy to expand $\text{Reg}_\alpha(t)$. $\square$

Lemma 3.3 *Lyapunov completion quality function $f_j'(S) = f_j(S) + V\sum_{i \in S} Z_i(t)$ for task j is submodular.*

Algorithm 3.2 UCB-based participant recruitment algorithm (LyaUCB-PRA) with fairness for task j in round t

---

Input: the covering set $\mathcal{N}_j^t$, the mean quality set $\{\hat{q}_i\}_i$ and the recruiting number set $\{n_i\}_i$, control parameter V, and queues $\mathbf{Z}(t)$

Output: the recruiting set $\mathcal{P}_j^t$

1: **if** $n_i = 0, \exists i \in \mathcal{N}_j^t$ **then**
2:      Randomly choose a subset containing user i: $\mathcal{P}_j^t \subseteq \mathcal{N}_j^t, |\mathcal{P}_j^t| \leq b_j^t$
3: **else**
4:      Calculate $\bar{q}_i, \forall i \in \mathcal{N}_j^t$ according to Eq. (3.8)
5:      Calculate Lyapunov UCB $H_i = V\bar{q}_i + Z_i(t), \forall i \in \mathcal{N}_j^t$
6:      Based on H_i, determine completion quality function $f_j'(\cdot)$
7:      Let $\mathcal{P}_j^t \leftarrow \emptyset$
8:      **for** $i = 1$ **to** b_j^t **do**
9:          **if** $i \leq \lceil 0.21 b_j^t \rceil$ **then**
10:            $k \leftarrow 2(b_j^t - i + 1)$
11:          **else**
12:            $k \leftarrow b_j^t$
13:          **end if**
14:          Greedy to choose a subset $M \subseteq \mathcal{N}_j^t \backslash \mathcal{P}_j^t$ to maximize $\sum_{i \in M} \Delta(i|\mathcal{P}_j^t)$, where $\Delta(i|\mathcal{P}_j^{*t}) = f_j(\mathcal{P}_j^t \cup \{i\}) - f_j(\mathcal{P}_j^t), |M| = k$
15:          Randomly select a user i from M
16:          $\mathcal{P}_j^t \leftarrow \mathcal{P}_j^{*t} \cup \{i\}$
17:      **end for**
18: **end if**
19: Based on $\mathcal{P}_j^{*t}$ determine the recruiting decision $x_{i,j}^t$ for user i, $\forall i \in \mathcal{N}_j^t$
20: Update $\mathbf{Z}(t)$ based on Eq. (3.11)
21: Update $\hat{q}_i$ and n_i once receiving the sensing data from user $i \in \mathcal{N}$
22: **return** $\mathcal{P}_j^t$.

Based on Lemmas 3.2 and 3.3, we can propose a UCB-based participant recruitment algorithm with a fairness guarantee, called LyaUCB-PRA. The pseudo-code is presented in Algorithm 3.2.

LyaUCB-PRA is derived from UCB-PRA in Algorithm 3.1 by integrating fairness constraints. Thus, LyaUCB-PRA additionally inputs the control parameter V and queues $\mathbf{Z}(t)$. And the additional operations include lines 5, 6, 19, and 20 for ensuring fairness constraints. However, the operations will not increase the computation complexity, which is still $O(b_j^t|\mathcal{N}_j^t|)$.

3.3.1.3 Privacy-Preserving Integration

Although the noisy feedback can blur users' quality, our proposed UCB-PRA and LyaUCB-PRA can efficiently learn the unknown users' quality with sublinear regret as stated in the above subsections, which can result in privacy leakage. To avoid the potential privacy leakage issue and further protect users' privacy regarding quality information, we aim to incorporate a locally differential privacy framework into our UCB-PRA and LyaUCB-PRA. To protect ς-LDP for the participants' quality, we allow the trusted server to perturb the noisy feedback using the Laplacian mechanism. In the Laplacian mechanism, the trusted server samples a random variable δ_i from Laplacian distribution with parameter $1/\varsigma$, and then perturbs the noisy feedback of user ι with variable δ_i before sending it to the platform. The platform will receive the noisy feedback and obtain the perturbed quality score $\tilde{\beta}_{ij}^t = \beta_{ij}^t + \delta_i$. Finally, the platform updates $\hat{q}_i$ and n_i with $\tilde{\beta}_{ij}^t$ in order to perform UCB-PRA and LyaUCB-PRA. It is proved that the Laplacian mechanism protects ς-LDP [56].

Note that there is no need to modify UCB-PRA and LyaUCB-PRA for taking ς-LDP into consideration. The only modification on the incentive protocol is that the trusted server decides the privacy parameter $\varsigma > 0$ and generates corresponding Laplacian noises to blur users' quality feedback before sending the feedback to the platform. Due to the consideration of ς-LDP, it additionally introduces a regret-privacy tradeoff to our proposed UCB-PRA and LyaUCB-PRA. We have the following theorem to characterize the emerging tradeoff.

Theorem 3.3 *For privacy parameter $\varsigma > 0$, UCB-PRA can achieve following LDP-based regret,*

$$Reg_\alpha(T) \leq 93J\sqrt{bNT\ln(T)}/\varsigma + \alpha JN\ln(T)$$
$$+ (1 + \zeta(\frac{2\eta - 1}{\varsigma}))\alpha JN, \tag{3.13}$$

and LyaUCB-PRA can achieve the following time-average LDP-based regret,

$$\overline{Reg}_\alpha(T) \leq \frac{N}{2V} + 93J\sqrt{bN\frac{\ln(T)}{\varsigma^2 T}} + \alpha JN\frac{\ln(T)}{T}$$
$$+ \frac{(1 + \zeta(\frac{2\eta-1}{\varsigma}))\alpha JN}{T}. \tag{3.14}$$

Proof When the quality scores are perturbed before participants submit her collected data, the concentration inequality becomes $\mathbb{P}[|\hat{q}_i - q_i| \leq \sqrt{\frac{\eta\ln(t)}{32n_i}} + \sqrt{\frac{\eta\ln(t)}{\varsigma^2 n_i}}]$. Based on the Chernoff-Hoeffding inequality, we can prove a high probability event that bounds the regret related to privacy parameter ς.   □

3.3.2 Stage II: Users' Interest Set Determination

In Stage III, the algorithms require the information of the cover set $\mathcal{N}_j^t$ for each task $j \in \mathcal{M}_t$. The platform can derive the information of $\mathcal{N}_j^t$ when it receives the interest set information $S_i^t \subseteq M_t$ from each user i (i.e., $N_j^t = \{i|j \in S_i^t, \forall i \in N\}$). Note that each user i independently determines her interest set S_i^t to maximize her utility, which can be defined as the difference between the total rewards and the total cost. Let $y_{ij} \in \{0, 1\}$ be the binary variable indicating whether user i adds task j into her interest set S_i^t. Then, the utility of user i can be expressed as,

$$U_i(y) = \sum_{j \in M_t} y_{ij}(r_j^{*t} - c_{ij}), \qquad (3.15)$$

where r_j^{*t} is the reward of task j that is priced by the platform at Stage I and c_{ij} is the potential sensing cost for user-task pair (i, j). $y = (y_{ij})_{j \in M_t}$ is the decision vector of user i. In this chapter, we consider a simple scenario with a tractable cardinality constraint. Thus, the user utility maximization is formulated as,

$$\max \quad U_i(y)$$

$(S2UOPT)$

$$s.t. \quad \begin{cases} \sum_{j \in M_t} y_{ij} \leq \min\{\Gamma_i^t, d_i\}, \\ (y_{ij} \in \{0, 1\}). \end{cases}$$

The cardinality constraint restricts the size of interest set S_i^t where Γ_i^t is due to the current limited resources of user i and d_i is the maximum number of the tasks assigned to user i. The platform can determine d_i to guarantee extensive and diverse participation, which further ensures fairness indirectly. One way to determine d_i is related to the learned quality information in Stage III. Note that the problem $(S2UOPT)$ can be addressed in $O(|M_t| \log |M_t|)$, which can be implemented by adopting the quick sort algorithm in a decreasing-order π of $r_j^{*t} - c_{ij}$ and select the first $\min\{\Gamma_i^t, d_i\}$ tasks of π into S_i^t. Thus, the obtained solution for the problem $(S2UOPT)$ is an equilibrium decision in Stage II.

When users' optimal interest set is determined by themselves, the users' individual rationalities are achieved. Note that individual rationality is slightly different from the existing works in our setting. Although considering the traditional individual rationality constraint that the payment for the user should be no less than its cost could be explicit in the reward pricing formulation, this restriction requires all users' sensing costs on the platform side, which is impossible in the implementation.

3.3.3 Stage I: Platform's Reward Pricing Strategy

In Stage II, the problem $(S2UOPT)$ is dependent on the tasks' rewards information, which is priced by the platform in Stage I. According to the problem description in Sect. 3.2.6, the platform determines optimal reward r_j^{*t} for task $j \in M_t$ by solving

Fig. 3.3 Probability that a
task is covered on different
rewards

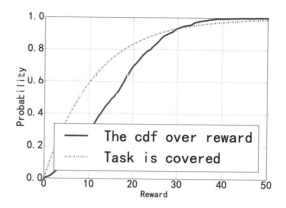

the following convex optimization,

$$\min \quad r_j^t$$

$$(S1POPT) \qquad s.t. \quad \begin{cases} 1 - e^{-\gamma_j^t r_j^t} \geq \frac{k_j^t}{N}, \\ b_j^t r_j^t \leq p_j^t, \\ (r_j^t \geq 0). \end{cases}$$

The first inequality is participation constraint and the second inequality is budget
constraint. Due to the first constraint containing convex functions, the feasible
domain is convex. Thus, problem $(S1POPT)$ is a convex optimization. According
to the KKT conditions, we can derive its closed form for its solution as follows,

$$r_j^{*t} = \min\{-\frac{1}{\gamma_j^t} \ln(1 - \frac{k_j^t}{N}), \frac{p_j^t}{b_j^t}\}. \tag{3.16}$$

From Eq. (3.16), we can see that the optimal reward r_j^{*t} is dependent on the total
payment p_j^t, the requirements for recruiting numbers (i.e., b_j^t and k_j^t) and the
hardness parameter γ_j^t of task j. And the solution r_j^{*t} for each task $j \in \mathcal{M}_t$ is
equilibrium reward in Stage I. We present the impact of rewards on participation
probability in Fig. 3.3.

3.4 Equilibrium Analysis

In this section, we present an equilibrium analysis of the proposed fair incentive
mechanism to show the strategy outcome of the MCS system under the fair proposed
incentive mechanism.

3.4.1 Strategy Performance in Stage III

Recall that in Stage III, we proposed a UCB-based participant recruitment algorithm with a fairness guarantee, i.e., LyaUCB-PRA as shown in Algorithm 3.2. With fairness constraint, we evaluate its optimality by regret-fairness tradeoff performance, which has the following theoretical result,

Theorem 3.4 *Given control parameter $V > 0$, there exists constants $B > 0, \epsilon > 0$ such that the overall queues' backlogs and α-regret in time-average in LyaUCB-PRA are satisfied the following inequalities,*

$$\limsup_{t \to \infty} \frac{1}{t} \sum_{\tau=0}^{t-1} \sum_{i=1}^{N} \mathbb{E}[|Z_i(t)|] \le \frac{B+V}{\epsilon}, \tag{3.17}$$

$$\overline{Reg}_\alpha(T) \le \frac{N}{2V} + 93J\sqrt{bN\frac{\ln(T)}{T}} + \alpha JN\frac{\ln(T)}{T} \\ + \frac{(1 + \zeta(2\eta - 1))\alpha JN}{T}. \tag{3.18}$$

Proof The proof is established on the drift-plus-penalty similar to Lemma 3.2 and UCB analysis like Theorem 3.1 over time slot $t = 1, 2, \ldots, T$. However, we can design a new concentration inequality proper for the combination in this setting. □

Theorem 3.4 demonstrates the regret-fairness tradeoff in a theoretical manner by a control parameter V. When V is larger, LyaUCB-PRA tends to minimize the regret to learn users' quality information, which will increase the backlogs of the queues $Z(t)$ to sacrifice the fairness constraints. When V is smaller, LyaUCB-PRA prioritizes the fairness constraints. Note that LyaUCB-PRA only outputs the recruiting decision for task j in round t, which does not make full use of arm dependency. The extension is similar to UCB-PRA-I.

3.4.2 Strategy Performance in Stage II

Recall that the strategy problem in Stage II is the problem $(S2UOPT)$. It should be noted that the problem $(S2UOPT)$ can be addressed in $O(|\mathcal{M}_t|\log|\mathcal{M}_t|)$, which can be implemented by adopting the quick sort algorithm in a decreasing-order π of $r_j^{*t} - c_{ij}$ and select the first $\min\{\Gamma_i^t, d_i\}$ tasks of π into $\mathcal{S}_i^t$. Thus, the obtained solution for the problem $(S2UOPT)$ is an equilibrium decision in Stage II. Therefore, we obtain an equilibrium strategy with a quick sort algorithm.

3.4.3   Strategy Performance in Stage I

Recall that the strategy problem in Stage I is the problem $(S1POPT)$, which is a convex optimization. According to the KKT conditions, we can derive its closed-form for its solution as follows,

$$r_j^{*t} = \min\{-\frac{1}{\gamma_j^t} \ln(1 - \frac{k_j^t}{N}), \frac{p_j^t}{b_j^t}\}.$$

Therefore, the strategy r_j^{*t} is an equilibrium strategy.

3.5   Performance Evaluation

In this section, we conduct simulations to evaluate the performance of the proposed incentive mechanism. The simulations are run in a computer with settings: Intel(R) Core(TM) i7-8700 CPU @3.40 GHz and 16 GB RAM, which are implemented in Python. The three-stage Stackelberg game interactions for the MCS scenario are simulated by synthetic trace data with $T = 1000$ rounds. The parameters are set according to Table 3.2 where we simulate $N = 100$ users and $M = 30$ requesters. We uniformly assign the true quality q_i of each user i over the interval $[0, 1.0]$ at random. For each user i, we assign the fairness guarantee threshold F_i with the UCB estimate of the quality $\bar{q}_i$ derived using Eq. (3.8), which ensures that the high-quality users are selected frequently. The task budget Γ_i^t is set to be the random value over the interval $[1, 10]$ and the maximum number d_i of the tasks assigned to user i is set to 5. In each round t, requester j provides the platform with a payment p_j^t (reward budget) that follows the uniform distribution (i.e., $\sim U(1, 40.0)$) over the interval $[1, 40.0]$. The recruiting budget b_j^t is randomly set over the interval $[2, 4]$ according to the uniform distribution. The covering number k_j^t is set to be 2, which means at least 2 users are selected to perform task j. The hardness parameter γ_j^t of task j is set to be 0.15. The sensing cost for user i to perform task j is generated according to the Gaussian distribution $N(15, 0.2)$ with mean 15 and variance 0.2. The exploration factor η is set to be 1.5 while the fairness scaling factor θ is set to

Table 3.2 Key parameters in simulations. $U(\cdot, \cdot)$ and $N(\cdot, \cdot)$ resp. refer to the uniform and normal distributions.

Parameters	Values	Parameters	Values
N	100	M	30
Γ_i^t	$\sim U(1, 10)$	d_i	5
q_i	$\sim U(0, 1.0)$	c_{ij}	$\sim N(15, 0.2)$
p_j^t	$\sim U(1, 40.0)$	T	1000
γ_j^t	0.15	b_j^t	$\sim U(2, 4)$
η	1.5	θ	0.2

be 0.2. We use Dixit-Stiglitz function $f_j(\mathcal{P}^t_j) = (\sum_{i \in \mathcal{P}^t_j} [q_i]^p)^{1/p}$ [16] with $p \geq 1$ to model the task completion function where different tasks have different p value.

3.5.1   Benchmarks and Metrics

To evaluate the regret performance, we compare our proposed UCB-PRA and LyaUCB-PRA with the following benchmarks,

- ε-**Greedy**: In round t, with probability ε the algorithm recruits the users who are less recruited in the past while with probability $1 - \varepsilon$, the algorithm recruits the optimal users according to the learned quality information.
- **Random**: In round t, the algorithm randomly select b^t_j users from $\mathcal{N}^t_j$ to perform task j.

To evaluate the fairness guarantee, we use two metrics: time average queues backlog $\overline{Z} = \frac{1}{t}\sum_{\tau=0}^{t-1}\sum_{i=1}^{N}\mathbb{E}[|Z_i(t)|]$ and the time-average recruiting rate $\overline{x}_i$ for each user $i \in \mathcal{N}$. $\overline{Z}$ is the implicit indicator while $\overline{x}_i$ is the explicit metric of the algorithm LyaUCB-PRA. To evaluate the privacy-preserving performance, we choose the privacy variance $\varsigma = 0.2, 0.5, 1.0, 2.0$. When ς is sufficiently large, our participant recruitment algorithm tends to sacrifice privacy to learn users' sensing qualities, and vice versa.

3.5.2   Evaluation Results

In Fig. 3.4, we present the regret performance of our first proposed UCB-PRA that only learns the users' sensing qualities without fairness and privacy guarantee. We can see that our UCB-PRA achieves a sublinear regret when T goes large. Moreover, UCB-PRA outperforms the benchmarks (**Random** and ε-**Greedy**) in terms of learning the users' sensing qualities information. ε-**Greedy** has an advantage under

Fig. 3.4 Regret performance of UCB-PRA

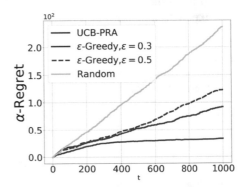

Fig. 3.5 UCB-PRA with
LDP, vary ς

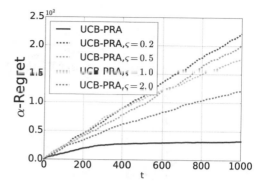

Fig. 3.6 Regret performance
of LyaUCB-PRA

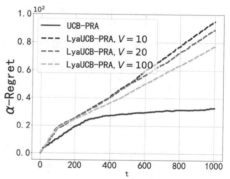

a small T case due to the explicit exploitation with $1 - \varepsilon$ probability at each round. However, our UCB-PRA has the asymptomatic optimal guarantee under a large T case.

In Fig. 3.5, we show the privacy-preserving performance of our UCB-PRA under difference privacy variance ς. When ς is smaller, the regret incurred by UCB-PRA becomes larger. This is because UCB-PRA protects users' privacy under a small ς case and incurs more privacy costs to learn the users' sensing qualities, which demonstrates a regret-privacy tradeoff coinciding with the result in Theorem 3.3.

In Figs. 3.6 and 3.7, we evaluate the fairness performance of LyaUCB-PRA under three fairness control parameter $V = 10, 20, 100$. We also compare the fairness performance with UCB-PRA which fails to guarantee the fairness of participant recruitment. Our considered performance metrics are regret and queues backlogs. The results in Fig. 3.6 show that (1) it incurs more regret to guarantee the fairness of participant recruitment; (2) the larger the fairness parameter V is, the less regret LyaUCB-PRA incurs. However, LyaUCB-PRA maintains larger queues backlogs under a large V case, as shown in Fig. 3.7. The larger queues backlogs indicate poor fairness performance as the recruiting decisions are only distributed the fewer users. The results of Figs. 3.6 and 3.7 jointly verify the regret-fairness tradeoff in Theorem 3.4.

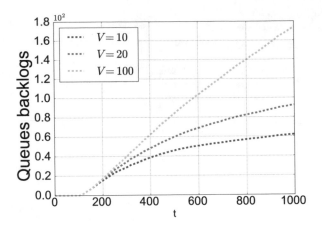

Fig. 3.7 Queues backlogs in LyaUCB-PRA, vary V

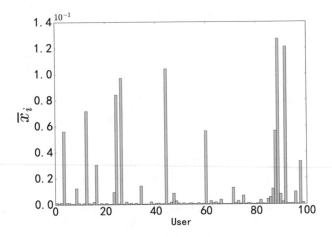

Fig. 3.8 Time-average rate $\overline{x}_i$ of UCB-PRA

To explicitly demonstrate the fairness performance, we plot the time-average recruiting rate $\overline{x}_i$ for all users in Figs. 3.8, 3.9, 3.10, and 3.11 when conducting participant recruitment using UCB-PRA and LyaUCB-PRA. It can be seen that the $\overline{x}_i$ in Fig. 3.8 is loosely distributed over different users because of no fairness guarantee in UCB-PRA. From Figs. 3.9, 3.10, and 3.11, $\overline{x}_i$ is uniformly distributed over different users due to the fairness guarantee of LyaUCB-PRA.

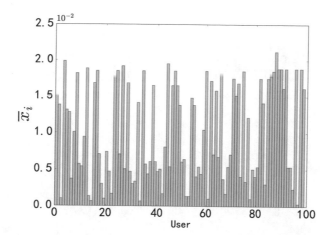

Fig. 3.9 Rate $\bar{x}_i$ of LyaUCB-PRA, $V = 10$

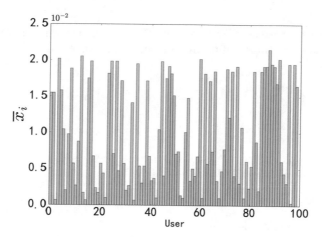

Fig. 3.10 Rate $\bar{x}_i$ of LyaUCB-PRA, $V = 20$

3.6 Conclusion

In this chapter, we propose a three-stage Stackelberg incentive mechanism for the MCS system with long-term interactions. Moreover, we propose participant recruitment algorithms (PRA) handle the unknown users' sensing qualities while guaranteeing fairness and privacy for users. We derive a sublinear regret bound $O(\alpha\sqrt{bNT \log T})$ for an online learning-based PRA. We also rigorously derive regret bound when ensuring fairness and protecting ς-differential privacy with $O(\alpha N \frac{\log T}{T})$ and $O(\alpha \varsigma^{-1}\sqrt{bNT \log T})$, respectively. Besides, we also address the problems of users and platform utility maximizations in different stages' interactions, which generate the Stackelberg equilibrium decisions. Finally, we

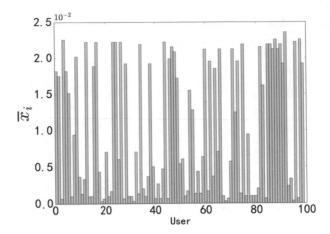

Fig. 3.11 Rate $\bar{x}_i$ of LyaUCB-PRA, $V = 100$

evaluate the performance of our proposed PRAs using simulations. In the future, we will consider the impact of the user's context and extend the simulations to real-world traces.

References

1. Liu, Y., Kong, L., Chen, G.: Data-oriented mobile crowdsensing: a comprehensive survey. IEEE Commun. Surv. Tuts. **21**(3), 2849–2885 (2019)
2. Guo, B., Liu, Y., Wang, L., Li, V.O.K., Jacqueline, C.K., Yu, Z.: Task allocation in spatial crowdsourcing: current state and future directions. IEEE Internet Things J. **5**, 1749–1764 (2018)
3. Bardin, N.: Waze (2008). https://www.waze.com
4. Chen, H., Li, F., Hei, X., Wang, Y.: Crowdx: enhancing automatic construction of indoor floorplan with opportunistic encounters. Proc. ACM Interact. Mobile Wearable Ubiquit. Technol. **2**(4), 1–21 (2018)
5. Wang, L., Zhang, D., Pathak, A., Chen, C., Xiong, H., Yang, D., Wang, Y.: CCS-TA: quality-guaranteed online task allocation in compressive crowdsensing. In: ACM International Joint Conference on Pervasive and Ubiquitous Computing (UbiComp), pp. 683–694 (2015)
6. Li, J., Cai, Z., Yan, M., Li, Y.: Using crowdsourced data in location-based social networks to explore influence maximization. In: IEEE International Conference on Computer Communications (INFOCOM), pp. 1–9 (2016)
7. Li, H., Li, T., Li, F., Yang, S., Wang, Y.: Multi-expertise aware participant selection in mobile crowd sensing via online learning. In: IEEE International Conference on Mobile Ad Hoc and Sensor Systems (MASS), pp. 433–441 (2018)
8. Liu, W., Yang, Y., Wang, E., Wu, J.: Dynamic user recruitment with truthful pricing for mobile crowdsensing. In: IEEE International Conference on Computer Communications (INFOCOM) (2020)
9. Gao, G., Wu, J., Xiao, M., Chen, G.: Combinatorial multi-armed bandit based unknown worker recruitment in heterogeneous crowdsensing. In: IEEE International Conference on Computer Communications (INFOCOM) (2020)

10. Gao, L., Hou, F., Huang, J.: Providing long-term participation incentive in participatory sensing. In: IEEE International Conference on Computer Communications (INFOCOM) (2015)
11. Han, K., Huang, H., Luo, J.: Quality-aware pricing for mobile crowdsensing. IEEE/ACM Trans. Netw. **26**(4), 1728–1741 (2018)
12. Lin, J., Yang, D., Li, M., Xu, J., Xue, G.: Bidguard: a framework for privacy-preserving crowdsensing incentive mechanisms. In: IEEE Conference on Communications and Network Security (CNS), pp. 145–153 (2016)
13. Li, T., Qiu, Z., Cao, L., Li, H., Guo, Z., Li, F., Shi, X., Wang, Y.: Participant grouping for privacy preservation in mobile crowdsensing over hierarchical edge clouds. In: IEEE Proceedings of the 37th International Performance Computing and Communications Conference (IPCCC), pp. 1–8. IEEE, Piscataway (2018)
14. Zhang, C., Zhu, L., Xu, C., Liu, X., Sharif, K.: Reliable and privacy-preserving truth discovery for mobile crowdsensing systems. IEEE Trans. Depend. Secur. Comput. **18**, 1245–1260 (2019)
15. Liu, Y., Wang, H., Peng, M., Guan, J., Wang, Y.: An incentive mechanism for privacy-preserving crowdsensing via deep reinforcement learning. IEEE Internet Things J. **8**(10), 8616–8631 (2020)
16. Chen, L., Xu, J., Lu, Z.: Contextual combinatorial multi-armed bandits with volatile arms and submodular reward. In: Advances in Neural Information Processing Systems (NeurIPS), pp. 3247–3256 (2018)
17. Jin, H., Su, L., Xiao, H., Nahrstedt, K.: Inception: incentivizing privacy-preserving data aggregation for mobile crowd sensing systems. In: ACM International Symposium on Mobile Ad Hoc Networking and Computing (MobiHoc), pp. 341–350 (2016)
18. Jin, H., Su, L., Ding, B., Nahrstedt, K., Borisov, N.: Enabling privacy-preserving incentives for mobile crowd sensing systems. In: IEEE International Conference on Distributed Computing Systems (ICDCS), pp. 344–353. IEEE, Piscataway (2016)
19. Wang, X., Liu, Z., Tian, X., Gan, X., Guan, Y., Wang, X.: Incentivizing crowdsensing with location-privacy preserving. IEEE Trans. Wireless Commun. **16**(10), 6940–6952 (2017)
20. Zhang, X., Liang, L., Luo, C., Cheng, L.: Privacy-preserving incentive mechanisms for mobile crowdsensing. IEEE Pervasive Comput. **17**(3), 47–57 (2018)
21. Wang, Z., Pang, X., Hu, J., Liu, W., Wang, Q., Li, Y., Chen, H.: When mobile crowdsensing meets privacy. IEEE Commun. Mag. **57**(9), 72–78 (2019)
22. Wang, Z., Li, J., Hu, J., Ren, J., Li, Z., Li, Y.: Towards privacy-preserving incentive for mobile crowdsensing under an untrusted platform. In: IEEE International Conference on Computer Communications (INFOCOM), pp. 2053–2061 (2019)
23. Zhao, B., Tang, S., Liu, X., Zhang, X.: Pace: privacy-preserving and quality-aware incentive mechanism for mobile crowdsensing. IEEE Trans. Mobile Comput. **20**(5), 1924–1939 (2020)
24. Wang, L., Cao, Z., Zhou, P., Zhao, X.: Towards a smart privacy-preserving incentive mechanism for vehicular crowd sensing. Securi. Commun. Netw. **2021**, 5580089 (2021)
25. Li, F., Liu, J., Ji, B.: Combinatorial sleeping bandits with fairness constraints. In: IEEE International Conference on Computer Communications (INFOCOM), pp. 1702–1710 (2019)
26. Xiao, M., Gao, G., Wu, J., Zhang, S., Huang, L.: Privacy-preserving user recruitment protocol for mobile crowdsensing. IEEE/ACM Trans. Netw. **28**(2), 519–532 (2020)
27. Yu, H., Iosifidisy, S., Biying, L., Huang, J.: Market your venue with mobile applications: collaboration of online and offline businesses. In: IEEE International Conference on Computer Communications (INFOCOM) (2018)
28. Chen, W., Wang, Y., Yuan, Y.: Combinatorial multi-armed bandit: general framework and applications. In: Proceedings of the 30th International Conference on Machine Learning (ICML), pp. 151–159. ACM, New York (2013)
29. Chen, W., Hu, W., Li, F., Li, J., Liu, Y., Lu, P.: Combinatorial multi-armed bandit with general reward functions. In: Advances in Neural Information Processing Systems (NeurIPS), pp. 1659–1667. MIT Press, Cambridge (2016)

30. Xiong, H., Zhang, D., Chen, G., Wang, L., Gauthier, V., Barnes, L.E.: ICrowd: near-optimal task allocation for piggyback crowdsensing. IEEE Trans. Mobile Comput. **15**(8), 2010–2022 (2015)
31. Liu, Y., Guo, B., Wang, Y., Wu, W., Yu, Z., Zhang, D.: Taskme: multi-task allocation in mobile crowd sensing. In: ACM International Joint Conference on Pervasive and Ubiquitous Computing (UbiComp), pp. 403–414 (2016)
32. Tao, X., Song, W.: Location-dependent task allocation for mobile crowdsensing with clustering effect. IEEE Internet Things J. **6**(1), 1029–1045 (2018)
33. Zhou, P., Chen, W., Ji, S., Jiang, H., Yu, L., Wu, D.: Privacy-preserving online task allocation in edge-computing-enabled massive crowdsensing. IEEE Internet Things J. **6**(5), 7773–7787 (2019)
34. Karaliopoulos, M., Telelis, O., Koutsopoulos, I.: User recruitment for mobile crowdsensing over opportunistic networks. In: 2015 IEEE Conference on Computer Communications (INFOCOM), pp. 2254–2262. IEEE, Piscataway (2015)
35. Wang, E., Yang, Y., Wu, J., Liu, W., Wang, X.: An efficient prediction-based user recruitment for mobile crowdsensing. IEEE Trans. Mobile Comput. **17**(1), 16–28 (2017)
36. Yang, D., Xue, G., Fang, X., Tang, J.: Incentive mechanisms for crowdsensing: crowdsourcing with smartphones. IEEE/ACM Trans. Netw. **24**(3), 1732–1744 (2016)
37. Cheung, M.H., Hou, F., Huang, J.: Make a difference: diversity-driven social mobile crowd-sensing. In IEEE International Conference on Computer Communications (INFOCOM) (2017)
38. Xiao, L., Li, Y., Han, G., Dai, H., Poor, H.V.: A secure mobile crowdsensing game with deep reinforcement learning. *IEEE Trans. Inf. Forensics Secur.* **13**(1), 35–47 (2018)
39. Chen, Y., Li, B., Zhang, Q.: Incentivizing crowdsourcing systems with network effects. In: IEEE International Conference on Computer Communications (INFOCOM) (2016)
40. Zhang, Y., Gu, Y., Pan, M., Tran, N.H., Dawy, Z., Han, Z.: Multi-dimensional incentive mechanism in mobile crowdsourcing with moral hazard. IEEE Trans. Mobile Comput. **17**(3), 604–616 (2018)
41. Jin, H., Guo, H., Su, L., Nahrstedt, K., Wang, X.: Dynamic task pricing in multi-requester mobile crowd sensing with markov correlated equilibrium. In: IEEE International Conference on Computer Communications (INFOCOM), pp. 1063–1071 (2019)
42. Duan, Z., Li, W., Cai, Z.: Distributed auctions for task assignment and scheduling in mobile crowdsensing systems. In: IEEE International Conference on Distributed Computing Systems (ICDCS), pp. 635–644 (2017)
43. Duan, Z., Li, W., Zheng, X., Cai, Z.: Mutual-preference driven truthful auction mechanism in mobile crowdsensing. In: IEEE International Conference on Distributed Computing Systems (ICDCS), pp. 1233–1242 (2019)
44. Cai, Z., Duan, Z., Li, W.: Exploiting multi-dimensional task diversity in distributed auctions for mobile crowdsensing. IEEE Trans. Mobile Comput. **20**(8), 2576–2591 (2020)
45. Zhan, Y., Xia, Y., Zhang, J., Li, T., Wang, Y.: An incentive mechanism design for mobile crowdsensing with demand uncertainties. Inf. Sci. **528**, 1–16 (2020)
46. Zhan, Y., Liu, C.H., Zhao, Y., Zhang, J., Tang, J.: Free market of multi-leader multi-follower mobile crowdsensing: an incentive mechanism design by deep reinforcement learning. IEEE Trans. Mobile Comput. **19**(10), 2316–2329 (2019)
47. Yu, H., Wei, E., Berry, R.A.: Monetizing mobile data via data rewards. IEEE J. Sel. Areas Commun. **38**, 782–792 (2020)
48. Neely, M.J.: Stochastic network optimization with application to communication and queueing systems. Synth. Lect. Commun. Netw. **3**(1), 1–211 (2010)
49. Bassily, R., Smith, A.: Local, private, efficient protocols for succinct histograms. In: Proceedings of the Forty-Seventh Annual ACM Symposium on Theory of Computing, pp. 127–135 (2015)
50. Chen, X., Zheng, K., Zhou, Z., Yang, Y., Chen, W., Wang, L.: (Locally) differentially private combinatorial semi-bandits. In: International Conference on Machine Learning (ICML) (2020)
51. Sun, L., Pang, H., Gao, L.: Joint sponsor scheduling in cellular and edge caching networks for mobile video delivery. IEEE Trans. Multimedia **20**(12), 3414–3427 (2018)

52. Krause, A., Golovin, D.: Submodular Function Maximization. Elsevier, Amsterdam (2014)
53. Buchbinder, N., Feldman, M., Naor, J.S., Schwartz, R.: Submodular maximization with cardinality constraints. In: Proceedings of the Twenty-Fifth Annual ACM-SIAM Symposium on Discrete Algorithms (SODA), pp. 1433–1452. SIAM, Philadelphia (2014)
54. Joulani, P., Gyorgy, A., Szepesvári, C.: Online learning under delayed feedback. In: International Conference on Machine Learning (ICML), pp. 1453–1461 (2013)
55. Bubeck, S., Cesa-Bianchi, N., et al.: Regret analysis of stochastic and nonstochastic multi-armed bandit problems. Found. Trends Mach. Learn. 5(1), 1–122 (2012)
56. Dwork, C., Roth, A., et al.: The algorithmic foundations of differential privacy. Found. Trends Theor. Comput. Sci. 9(3–4), 211–407 (2014)

Chapter 4
Collaborative Incentive Mechanism for Mobile Crowdsensing

Abstract In this chapter, we propose PTASIM, an incentive mechanism that explores cooperation with POI-tagging App for Mobile Edge Crowdsensing (MEC). PTASIM requests the App to tag some edges to be POI (Points-of-Interest), which further guides App users to perform tasks at that location. We further model the interactions of users, a platform, and an App by a three-stage decision process. The App first determines the POI-tagging price to maximize its payoff. Platform and users subsequently decide how to determine tasks reward and select edges to be tagged, and how to select the best task to perform, respectively. We analyze the optimal solution in those stages. Specifically, we prove greedy algorithm could provide the optimal solution for the platform's payoff maximization in polynomial time. The numerical results show that: (1) the cooperation with App brings long-term and sufficient participation; the optimal strategies reduce the platform's tasks cost as well as improve App's revenues.

Keywords Third-party collaboration · POI-tagging App · Participation rate guarantee · Three-stage decision process · Stackelberg game

4.1 Introduction

4.1.1 Motivations

As an emerging sensing paradigm in IoT, crowdsensing provides large-scale sensing services by recruiting mobile users with rich sensors to perform sensing tasks in the specific location [1, 2]. When the conventional crowdsensing architecture encounters the scalability problem due to the cloud-based implementation, the rising edge architecture and edge computing [3, 4] enables this challenge to be overcome and results in new architecture in IoT, called Mobile Edge Crowdsensing (MEC) [5, 6]. In MEC, the platform offloads data collection tasks [7] to mobile users via edges. Incentive mechanism is indispensable in crowdsensing to compensate users' selfishness, sensing cost and even privacy concerns [8]. However, existing works on incentive mechanisms focus on a single task or fixed location tasks (assuming

© The Author(s), under exclusive license to Springer Nature Singapore Pte Ltd. 2024
Y. Li et al., *Incentive Mechanism for Mobile Crowdsensing*, SpringerBriefs in Computer Science, https://doi.org/10.1007/978-981-99-6921-0_4

there are a large number of crowds in the corresponding location). These works do not consider users' heterogeneity (e.g., assuming users are willing to go to related locations and perform tasks as long as rewards are offered enough). As a result, they cannot be well extended to the scenario where the platform has a set of location-aware sensing tasks to be accomplished. In fact, it is more practical for the platform to take multiple location-aware tasks into account in crowdsensing [9, 10]. We find third-party cooperation is useful to complete multiple location-aware tasks. Unfortunately, none of the existing works shed light on the incentive mechanism design when considering third-party cooperation in crowdsensing. In MEC, part of the functionality is offloaded to the edges like user recruitment and the platform only focuses on receiving requests and distributing them to the edges.

To this end, we design a POI-Tagging App-aSsisted Incentive Mechanism (abbr. PTASIM), which exploits hybrid platforms' cooperation. POI-tagging Apps, especially augmented reality Apps, are more and more popular in our daily lives (e.g., Pokemon Go [11]). This kind of Apps brings its users intrinsic value (such as entertainment, and enjoyment), since it accommodates users' experience with the physical world leveraging emerging AR/VR technology. They are characterized by affecting App users' moving behaviors, even interacting actions. Thus, the key rationale of PTASIM is that a platform in the crowdsensing system turns to POI-tagging Apps and requests them to tag some locations of tasks as POI, where normal users are recruited costly or insufficient.

Figure 4.1 illustrates the insights of PTASIM proposed in this chapter. There are several location-aware tasks arriving in an i.i.d manner over time and 3 users (1 normal users and 2 App users). The platform receives sensing service requests

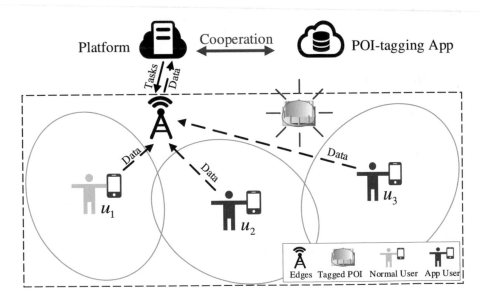

Fig. 4.1 PTASIM crowdsensing model

and sends tasks to the proximal edges. If the edges directly recruit users to perform tasks, then only normal users are available to be recruited as workers. Obviously, the participation rate[1] of the task is not enough for the crowdsensing system where one task needs to be done by multiple users such that data fusion can output fine results. When the platform cooperates with the POI-tagging App and requests the App to tag the location of the edge as POI, there are 3 users (1 normal user and 2 App users) to perform the task, which improves users participation in sensing tasks.

4.1.2   Challenges and Our Contributions

Despite the great potential of hybrid platform cooperation, designing PTASIM remains challenging because: (1) the tasks are location-aware and distributed differently. As a result, there is a conflict between different tasks for recruiting users. The platform has to recruit enough users for different tasks in the corresponding location; (2) users and the App usually are not willing to cooperate with the platform unless proper payments are offered. Each of them inclines to maximize its own payoff with the best strategy; (3) the decisions should be made instantaneously rather than in an offline manner. This will be more practical when handling large-scale location-aware task scenarios. The key issues are how to determine tasks rewards and how to select tasks to be tagged from the prospect of the platform as well as how to determine POI-tagging price from the prospect of the App.[2] In this chapter, we address these challenges by formulating PTASIM as a three-stage decision process and derive optimal incentive policy accordingly. We summarize the major contributions of this chapter as follows:

- We propose a novel incentive mechanism for hybrid platform cooperation. We study a multiple location-aware tasks incentive mechanism in crowdsensing and integrating cooperation with POI-tagging App into incentive mechanism design.
- We model PTASIM designing problem as a three-stage decision process and further analyze the optimal solution in each stage for users, platform, and App, respectively. More importantly, we propose a greedy algorithm to address the platform's payoff maximization and prove its optimality using Matroid.
- We conduct simulations to validate our proposed PTASIM. Numerical results show that cooperation with POI-tagging App improves the participation rate and reduces tasks cost for the platform while increasing App's revenues.

---

[1] We assume participant recruitment happens in a specific short period so that the total number of users is fixed during that time. Therefore, we apply the participant ratio as the requirement for the rest of this chapter, which is the same as the required number of users.

[2] In this chapter, we assume POI-tagging App easily guides its users to perform tasks. And the quality and capability of performing tasks for normal users and App users are the same.

4.1.3 Related Work

There are some existing works focusing on pure edge computing [12, 13] and privacy-preserving in crowdsensing [14–16]. In this chapter, we pay attention to designing an incentive mechanism for mobile edge crowdsensing.

MSensing game and MSensing Auction [17] design incentive mechanisms based on the Stackelberg game and auction for the platform-centric model and user-centric model, respectively. Lyapunov-based VCG auction [18] studies long-term participation incentives by combining the Lyapunov optimization framework and VCG auction. TBA, TOIM and TOIMAD [19] are online incentive mechanisms based on the reverse auction. Network effect-based incentive mechanism [20] explores the impact of intrinsic rewards resulting from network effect due to large participation in more efficient extrinsic rewards design. Diversity-driven and social effect-based reward mechanism [21] exploit users' diversity for sensing quality as well as social effect from users' social relationship for incentivizing users' participation. However, these works mainly focus on incentive mechanisms for a single task or specific time-dependent and location-aware tasks. In this chapter, we focus on designing a practical incentive mechanism for multiple different tasks in crowdsensing.

There are also some existing works that study multiple tasks for crowdsensing. Truthful incentive mechanisms [22] study SS-model, SM-model and MM-model involving cooperation and competition among the service providers based on auction with truthfulness. CENTURION [23] utilizes double auction to design an incentive mechanism stimulating the participation of both multiple data requesters and workers to complete multiple tasks. Taskme [24] formulates multi-task allocation as bi-objective optimization for FPMT and MPFT. MRA and MCA [25] are reliable task assignments for crowdsensing. Although these works focus on multi-task-oriented crowdsensing, they do not involve hybrid platform cooperation and exploit the collaboration with POI-tagging App. In this chapter, we design an incentive mechanism leveraging the cooperation with POI-tagging App for efficient participant recruitment and effective rewards offering.

In terms of research approach, three-stage Stackelberg game (or decision process) [21, 26–29] are used to model pricing problems in different areas. POI-based two-part pricing [26] involves POI-based collaboration, their model can derive closed-form solutions due to scenario-specific reasons. However, their methods cannot be directly extended to the incentive mechanism design in crowdsensing, since the payoff definitions are different.

4.2 Game Modeling

In this section, we first describe the strategies and attributes for decision-makers (users, platform, and App) in location-aware MCS. And then their payoffs are formulated based on their strategies, respectively. The mathematical notations in this chapter are summarized in Table 4.1.

4.2.1 User

A user u_j in location-aware MCS is characterized by a tuple (L, f_{app}). Here, f_{app} is a boolean variable indicating whether the user uses the App and L is the remaining battery level of the user's smart device. We assume L is uniformly distributed in $[0, 1]$. Let L_{th} denote the threshold that the user will not perform sensing task or use App if its L is lower than L_{th}. We use $\kappa \in [0, 1]$ to represent the proportion of

Table 4.1 Notations used in Chap. 4

Notations		Meanings
Users	$N, \mathcal{U}$	The number of users, user set
	f_{app}	Boolean variable to indicate App user
	L	Remaining battery level of mobile devices
	κ	The proportion of App users
	$\Phi_{ij}(\cdot)$	The payoff task t_i brought to user u_j
	E_i^δ	Congestion effect with factor δ about task t_i
	E_i^θ	Network effect with factor θ about task t_i
	c_{ij}	Cost of user u_j for performing task t_i
	d_{ij}	Decision variable of user u_j to t_i
Platform	$M, \mathcal{TS}$	The number of tasks, task set
	γ_i	The revenues of task t_i
	ρ_i	The rewards priced for task t_i
	n_i	Task t_i's required number of users
	I_i	The measurement for network infrastructure
	r_i	Boolean variable to indicate POI-tagging
	$\widetilde{x}_i(r_i)$	Total users participation rate for task t_i
	$\widetilde{y}_i(r_i)$	App users participation rate for task t_i
	Φ_p	The payoff of platform
App	V	Utility of using App
	P_{app}	POI-tagging price App charges
	a	App's unit profit for advertisement revenues
	C_{tag}	The cost for App to tag POI
	Φ_{app}	The payoff of App

App users. We further consider a user set $\mathcal{U} = \{u_1, u_2, \ldots, u_N\}$. Normal users will select a task with the maximal payoff. App users will go to the POI-tagged location and perform the selected task while using App. Although there possibly exist some users who are both normal users and App users, we account for these users for App users since the number of payoffs of App users is greater than the number of payoffs of normal users as described later.

4.2.2   Platform

The platform in MCS has M location-aware sensing tasks. The platform either directly recruits users to perform sensing tasks in POI by offering incentive payments or cooperatively requests the App to tag POI for sensing tasks. Suppose that the platform's task set is $\mathcal{TS} = \{t_1, t_2, \ldots, t_M\}$. Each task t_i is represented by a tuple $(\gamma_i, \rho_i, n_i, I_i)$. γ_i is the revenue generated to the platform when task t_i is completed. ρ_i is the reward of t_i priced by the platform. n_i is the required number of participants to perform task t_i due to sensed data fusion for high quality. I_i is network infrastructure status (i.e., Wi-Fi, the signal strength of LTE) related to POI of t_i, and $I_i \in [0, 1]$. If I_i is close to 1, it indicates a good network infrastructure status. This can be measured from the number of infrastructures (LTE, Wi-Fi, D2D) and available bandwidth. The platform has to determine the sensing rewards as an incentive for users while selecting which tasks to be tagged.

4.2.3   App

We consider a POI-tagging App (such as Pokemon Go, and Snapchat). This kind of Apps can tag some locations as POIs and offer its users the utility of V when used. Moreover, the App can cooperate with the MCS platform and help the platform to tag POI for sensing task t_i and charge P_{app}, which is the POI-tagging price. In this chapter, we assume the App is free for users, which conforms to the practical scenario [26]. At first, the App should make a decision on P_{app} to maximize its payoff, which in turn determines how the platform cooperates with the App. Actually, the platform does not need to deploy this kind of POI-tagging Apps. Contrarily, the platform can utilize the cooperation with off-the-shelf Apps like Pokemon Go, and Snapchat in order to focus on providing sensing services and achieve cost-efficient goals.

4.2.4 Payoff Definition

4.2.4.1 Users' Payoff

When user u_j decides to perform sensing task t_i, its utility is $\rho_i - c_{ij}$. Here c_{ij} is the cost the user j suffers when he/she performs t_i (such as the transportation cost for u_j to go to the location of t_i) [26]. Without loss of generality, we assume c_{ij} is uniformly distributed in $[0, c_{max}]$. Moreover, if the App tags task t_i as POI, its user u_j will obtain utility V due to the entertainment generated by the App. Boolean variable $r_i = 1$ represents that the platform requests the App to tag sensing task t_i as POI, and 0 otherwise. We further consider the impact of network effect and congestion effect[3] [20, 28] on the utility of the user, and they are modeled by two variables: E_i^θ and E_i^δ, respectively. δ is the congestion effect coefficient, which is globally the same for all users including App users and normal users due to the common limited wireless resources like bandwidth. θ is the network effect coefficient. In practice, θ is heterogenous over different App users due to the diversity of intrinsic utility of different App users. The heterogeneity of θ will not change the result of Lemma 4.1 since it corresponds to a scenario for normal users. For Lemma 4.2, we can replace θ with θ_j for user u_j and analyze the effect of the heterogeneous network effect. Then, we need to derive a threshold θ_{th} by considering the net payoff in Eq. (4.1). For any $\theta_j < 0$, user j does not join to perform a task. Therefore, the result of Lemma 4.2 scales by factor $\frac{\theta_{th}}{\theta_{max}}$. However, it is complicated for analysis. For ease of exposition, we use the same θ which can reflect the bound to some extent. We leave the heterogeneity of θ to study in future work. Based on the above notations, if user u_j decides to perform task t_i (let boolean variable $d_{ij} = 1$ denote that user u_j perform task t_i, and $d_{ij} = 0$ otherwise), we define user's payoff with type (L, f_{app}) as follows:

$$\Phi_{ij}(L, f_{app}, \rho_i, r_i) = \begin{cases} (\rho_i - c_{ij} - E_i^\delta + r_i f_{app}(V + E_i^\theta)) \cdot d_{ij} & \text{if } L \geq L_{th}, \\ 0 & \text{otherwise.} \end{cases}$$

(4.1)

Here, E_i^θ and E_i^δ are calculated by Yu et al. [26]: $E_i^\theta = \theta \widetilde{y}_i(r_i)N$ and $E_i^\delta = \frac{\delta}{l_i} \widetilde{x}_i(r_i)N$, where $\widetilde{y}_i(r_i)$ is App users participation rate for task t_i and $\widetilde{x}_i(r_i)$ is total users participation rate for task t_i, which are a ratio of the number of users participating in task t_i to the number of all users in the system. $\widetilde{x}_i(r_i)$ and $\widetilde{y}_i(r_i)$ are very important to derive the optimal strategies for users, the platform and the App. At equilibrium, they converge to a fixed value [26]. In next section, we present how to calculate $\widetilde{x}_i(r_i)$ and $\widetilde{y}_i(r_i)$ in the case with $r_i = 0$ and $r_i = 1$. Note that

---

[3] In this chapter, network effect represents App's popularity and concentration which generates a positive externality for App users. Due to limited wireless bandwidth, we also consider the congestion effect which is a negative effect on all users simultaneously using the network to transfer the sensing data at the locations.

the assumptions of uniform distributions for L and c_{ij} do not affect the analytical results of PTASIM but only influence the simulation settings.

4.2.4.2   Platform's Payoff

Platform's utility depends on the number of completed tasks. Therefore, it should offer an incentive for users to perform sensing tasks in the corresponding location. Different from traditional incentive mechanism which unilaterally requires users to a specific location for performing sensing tasks [10, 17, 18, 20–24], our proposed POI-tagging incentive mechanism considers the cooperation with the POI-tagging App. The platform intuitively pays the App to tag the location related to sensing tasks and lets the users perform sensing tasks when users use the App. We also define platform strategy profile $p = (\rho_{t_1}, \rho_{t_2}, \ldots, \rho_{t_M})$ and $r = (r_1, r_2, \ldots, r_M)$. Under the POI-tagging price P_{app} announced by App and the equilibrium of user participation rate $\widetilde{x}_i(r_i)(\forall i)$, the payoff of the platform is defined as follows:

$$\Phi_p = \sum_{i=1}^{M} \mathbb{I}(\widetilde{x}_i(r_i) \geq \frac{n_i}{N})(\gamma_i - N\widetilde{x}_i(r_i)\rho_i) - r^T P_{app}, \qquad (4.2)$$

where $\mathbb{I}(\cdot)$ is an indicator function which equals to 1 if input inequality holds, and 0 otherwise. P_{app} is a vector where its elements are P_{app}. Clearly, if task t_i is performed by at least n_i users at equilibrium, $\mathbb{I}(\widetilde{x}_i(r_i) \geq \frac{n_i}{N}) = 1$.

4.2.4.3   App's Payoff

Under the POI-tagging price P_{app} announced by the App, the platform optimally chooses the number of tasks $|r^*(P_{app})|$ to be tagged. Therefore, App's revenues generated from helping platform tag POI for $|r^*(P_{app})|$ tasks is $P_{app}^T r^*(P_{app})$. In addition, the App can advertise to users for more revenue. Let $aN\kappa$ denote the advertisement revenues of the App, where a is the unit profit for advertisement revenues. Obviously, the App's payoff consists of POI-tagging revenues and advertisement revenues, thus denoting below:

$$\Phi_{app} = P_{app}^T r^*(P_{app}) + aN\kappa \qquad (4.3)$$

In our proposed incentive mechanism, App chooses the optimal POI-tagging price P_{app} to maximize its payoff by considering the platform's decision on $r^*(P_{app})$ and extra advertisement revenues.

4.2.5   Three-Stage Decision Process

Suppose that the platform ignores the cooperation, the users participation rate is $1 - \kappa$ in Stackelberg Equilibrium. The Price of Anarchy (PoA) [30], which is used to measure the inefficiency of incentive mechanism, is, therefore, $\frac{1}{1-\kappa}$. The larger κ is, the higher PoA is. Hence, the platform has more incentive to leverage POI-tagging cooperation under higher PoA situations.

We model the PTASIM as a three-stage decision process (three-stage Stackelberg game). The App first determines POI-tagging price P_{app}, and then the platform strategizes how much incentive reward to be offered and which tasks to be tagged (i.e., p and r) in Stage II. Finally, users with different types select one task to perform and go to its corresponding location. For ease of analysis, we present the optimal strategies for different roles in the different stages using backward induction in the following section [26–28, 31].

Specially, we illustrate the existence of equilibrium with respect to participation rate under the different cases (i.e. $r_i = 0$ and $r_i = 1$) in Sect. 4.3.1. The existence of the equilibrium allows the platform to predict the decision of the users and thus enables the platform to calculate the optimum value of p and r. Therefore, we derive the equilibrium of participation rate in Sect. 4.3.1 for any given $p \succeq 0$ and analyze the optimal strategy of p and r in Sect. 4.3.2, both of which forms Stackelberg Equilibrium [17].

4.3   Detailed Design with Equilibrium Analysis

4.3.1   Stage III: Tasks Allocation

In this section, we analyze the optimal strategies for users to select the best task. Leveraging the three-stage decision process, we derive users' decisions by considering the given p and POI-tagging decision vector r. Due to all users' best strategies, we further analyze the task's participation rate under cases with $r_i = 0$ and $r_i = 1$ (i.e., whether to tag task t_i or not). We can regard the case without tagging task t_i as a situation where its participation rate is so sufficient that the platform can directly recruit users, otherwise, the platform will turn to App for POI-tagging recruitment.

Clearly, both normal users ($f_{app} = 0$) and App users ($f_{app} = 1$) would like to select the best task which maximizes its payoff. We extend $d_{ij}^* = \arg\max_{d_{ij}} \Phi_{ij}(L, f_{app}, \rho_i, r_i)$ as follows:

$$(UPM) \quad \max \sum_{i=1}^{M} \Phi_{ij}(L, f_{app}, \rho_i, r_i)$$

$$s.t. \quad \begin{cases} \sum_{i=1}^{M} d_{ij} = 1, & (i = 1, 2, \cdots, M) \\ d_{ij} \in \{0, 1\}, & (i = 1, 2, \cdots, M) \end{cases}$$

The first constraint in UPM ensures each user can only perform one sensing task which brings it optimal payoff. Solving UPM, we can derive the best strategy for user u_j where u_j selects task t_i if and only if $d_{ij}^* = 1$. Solving UPM is to easily enumerate all tasks and derive the maximum one. Analyzing the best strategy for a single user is helpful for predicting the total users' behavior (i.e., $\widetilde{x}_i(r_i)$ and $\widetilde{y}_i(r_i)$). Obviously, if $L < L_{th}$, then user u_j has no interest in performing any task $t_i \in \mathcal{TS}$ due to lack of enough battery level. Under this circumstance, the solution to UMP is $d_{ij}^* = 0(\forall t_i)$. To derive a significantly far-reaching solution, we consider the general situation where $L \geq L_{th}$. The solution under sufficient energy situation ($L \geq L_{th}$) is concluded in the following propositions. The propositions cover the cases where $r_i = 0$ and $r_i = 1$.

4.3.1.1   The Case with $r_i = 0$

$r_i = 0$ means that platform directly recruits users for task t_i by offering rewards ρ_i. Let $\boldsymbol{p}_{-i}$, $\boldsymbol{c}_{-i,j}$ and $\boldsymbol{E}_{-i}^{\delta}$ denote that the tasks rewards vector, tasks to user u_j cost vector and congestion effect vector in tasks location excluding task t_i. We have the following proposition to show the solution to (UMP) under the case of directly recruiting.

Proposition 4.1 (Directly Recruiting) *For a task t_i where platform directly recruits users to perform by offering ρ_i, the best choice d_{ij}^* of user u_j with respect to the problem UPM is derived as*

$$d_{ij}^* = \begin{cases} 1 \; if \; \rho_i - c_{ij} - E_i^{\delta} \geq max\{\boldsymbol{p}_{-i} - \boldsymbol{c}_{-i,j} - \boldsymbol{E}_{-i}^{\delta}\}, \\ 0 \; otherwise. \end{cases} \tag{4.4}$$

When the condition in (4.4) holds in the form of equality, user u_j can select a task t_i with the minimal index or randomly select any task that brings the same utility. The rationale behind Proposition 4.1 is intuitive as we can see that user u_j seeks to select a task $t_i \in \mathcal{TS}$ which can bring it optimal revenues when the platform directly recruits it by announcing tasks rewards $\boldsymbol{p}$. There exists only task t_i to user u_j which satisfy $d_{ij}^* = 1$ because in most cases, there is only one task generating maximal utility to users while we have operator $H(\cdot)$ to handle the tie situation. Therefore, the solution from Proposition 4.1 conforms to the constraints of UPM.

Under the tasks rewarding p, there are a fraction of users incentivized to the locations corresponding to tasks which reflect in $\widetilde{x}_i(r_i)$ and $\widetilde{y}_i(r_i)$ ($\forall i$). Predicting $\widetilde{x}_i(r_i)$ and $\widetilde{y}_i(r_i)$ are important to the platform, as the platform has to calculate best tasks rewards p in the light of $\widetilde{x}_i(r_i)$ and $\widetilde{y}_i(r_i)$. Therefore, we have the following lemma to show the calculation of $\widetilde{x}_i(0)$ and $\widetilde{y}_i(0)$ corresponding to Proposition 4.1 under case with $r_i = 0$:

Lemma 4.1 *For a task t_i with $r_i = 0$ (i.e., platform directly recruits users for task t_i), the expected fraction of users to the location of task t_i at equilibrium is*

$$\begin{cases} \widetilde{x}_i(0) = \frac{(1-\kappa)\rho_i I_i}{(1-\kappa)N\delta + c_{max} I_i} \\ \widetilde{y}_i(0) = 0 \end{cases} \tag{4.5}$$

Proof It is easy to verify that $\widetilde{y}_i(0) = 0$ because no App users are expected to the location of task t_i which is not the POI tagged by App. For the calculation of $\widetilde{x}_i(0)$, we derive it by calculating the net user whose payoff is zero and combining the assumption that c_{ij} is uniformly distributed in $[0, c_{max}]$. □

4.3.1.2 The Case with $r_i = 1$

If the platform requests the App to tag POI for task t_i (i.e., $r_i = 1$), the App users may go to this POI to perform a task before using the App. On the other hand, the normal user ($f_{app} = 0$) may perform task t_i if t_i is considered as optimal task[4]. Under this case with $r_i = 1$, we conclude the optimal strategy d_{ij}^* of user u_j related UPM in the following proposition:

Proposition 4.2 (POI-Tagging Recruiting) *For a task t_i which is tagged as POI by App due to the request from the platform, the best choice d_{ij}^* of user u_j with respect to the problem UPM is derived as:*

$$d_{ij}^* = \begin{cases} 1 & \text{if } \rho_i - c_{ij} - E_i^\delta \geq max\{p_{-i} - c_{-i,j} - E_{-i}^\delta\} \\ & \text{and } f_{app} = 0, \\ 1 & \text{if } \rho_i - c_{ij} - E_i^\delta + E_i^\theta \geq \\ & \quad max\{p_{-i} - c_{-i,j} - E_{-i}^\delta + E_{-i}^\theta\} \\ & \text{and } f_{app} = 1, \\ 0 & \text{otherwise,} \end{cases} \tag{4.6}$$

---

[4] f_{app} is user's intrinsic attribute and independent of tagging tasks as POI.

As mentioned above, predicting $\widetilde{x}_i(r_i)$ and $\widetilde{y}_i(r_i)$ will facilitate platform to price p in the Stage II of formulated three-stage decision process. We also present how to calculate $\widetilde{x}_i(1)$ and $\widetilde{y}_i(1)$ under the case with $r_i = 1$ in following lemma:

Lemma 4.2 *For a task t_i with $r_i = 1$ (i.e., platform requests App to tag POI for task t_i in order to increase the number of users), the expected fraction of users to the location of task t_i at equilibrium is*

$$
\begin{cases}
\widetilde{x}_i(1) = \frac{I_i(\kappa(V+\rho_i)((1-\kappa)N\delta+c_{max}I_i)-(1-\kappa)I_i\rho_i(\kappa N\theta-c_{max}))}{(\kappa\delta N-I_i(\kappa\theta N-c_{max}))((1-\kappa)N\delta+c_{max}I_i)} \\
\widetilde{y}_i(1) = \frac{I_i\kappa((V+\rho_i)((1-\kappa)N\delta+c_{max}I_i)-(1-\kappa)N\delta\rho_i)}{(\kappa\delta N-I_i(\kappa\theta N-c_{max}))((1-\kappa)N\delta+c_{max}I_i)}
\end{cases} \tag{4.7}
$$

Proof Considering the total participation rate in the location of task t_i, we have:

$$
\widetilde{x}_i(0) + \widetilde{y}_i(1) = \widetilde{x}_i(1) \tag{4.8}
$$

Also considering the net App user whose payoff is zero in the location of task t_i, we have:

$$
\kappa \frac{\rho_i + V + \theta\widetilde{y}_i(1)N - \frac{\delta}{I_i}\widetilde{x}_i(1)N}{c_{max}} = \widetilde{y}_i(1). \tag{4.9}
$$

By solving Eq. (4.8) and Eq. (4.9) and combining with Lemma 4.1, we can verify that this lemma holds.   □

From Eq. (4.8), we can see that the participation rate for task t_i increases if the platform requests the App to tag POI for task t_i. This can improve the quality of returned sensing data while decreasing the incentive reward. Note that the participation rates with respect to $\widetilde{x}_i(r_i)$ and $\widetilde{y}_i(r_i)$ given in Lemma 4.1 or Lemma 4.2 are at equilibrium when expected the strategies of platform (ρ_i and r_i). The platform will use the equilibria participation rate to calculate the optimal incentive rewards p and select the best tagging tasks r, which will be analyzed in the next section.

4.3.2   Stage II: Incentive and Tagging Determination

In this section, we analyze the optimal strategies for the platform to determine tasks incentive rewards p^* and POI-tagging decision r^*. Intuitively, directly choosing p^* and r^* to maximize the platform's payoff in Eq. (4.2) is nontrivial due to the presence of indicator function. However, we find that platform's payoff maximization can be equivalently transformed into the following mathematical programming problem:

$$\min \ N p^T \tilde{x}(r) + r^T P_{app}$$

$$(PPM) \quad s.t. \quad \begin{cases} N p^T \tilde{x}(r) + r^T P_{app} \preceq \sum_{i=1}^{M} \gamma_i, \\ \tilde{x}_i(r_i) \geq \frac{n_i}{N}, \\ \rho_i \leq (1 - r_i)p_{th1}(I_i) + r_i\, p_{th2}(I_i), \\ p \succeq 0, \\ r_i \in \{0, 1\}, \\ (i = 1, 2, \cdots, M), \end{cases}$$

Here we define two thresholds as follows:

$$\begin{cases} p_{th1}(I_i) \triangleq \frac{(1-\kappa)N\delta + c_{max}I_i}{I_i} \\ p_{th2}(I_i) \triangleq \frac{I_i^2 c_{max}(c_{max} - V - \kappa\theta N) + I_i c_{max}\delta N}{c_{max}I_i^2} \\ \qquad\qquad + \frac{I_i N\delta(\kappa^2\theta N - \kappa\theta N + \kappa V - V)}{c_{max}I_i^2} \end{cases} \tag{4.10}$$

A simple approach to derive the solution of (PPM) is to use a brute force algorithm, which needs to check all the feasible solutions satisfying the constraints. This approach is called EXPPM and its complexity is $O(M^2 2^M)$.

4.3.2.1 Greedy Approach

Although PPM can be formulated by MIQP, the studied problem for the platform's decision luckily exists favorable properties facilitating to design polynomial algorithm in terms of calculating the optimal solution. In this section, we further analyze the problem from the perspective of Matroid [32] which induces the greedy algorithm to solve PPM exactly in polynomial time.

Definition 4.1 (Independent Tagging Set) For a set $S \subseteq TS$, if each task $t_i \in S$ is tagged with reward ρ_i satisfying the constraints in PPM and its objective in PPM is no more than $obj(0)$, S is called independent tagging set. Its size is called a rank. Here, $obj(\cdot)$ represents the objective of PPM under tagging vector r.

With the definition of an independent tagging set, we can reformulate PPM as a matroid $M = (TS, I)$, where I is a non-empty collection of an independent subset of TS. M has following properties:

- (transitivity) Every subset of $I \in \mathcal{I}$ is independent.
- (commutativity) If $A \in \mathcal{I}$, $B \in \mathcal{I}$ and $|A| < |B|$, then there exists an element $x \in B - A$ satisfying $A \cup \{x\} \in \mathcal{I}$.

hence, PPM is equivalently solved when we find an independent subset with maximal rank in $\mathcal{M}$. We have the following lemmas to support the fact that searching the maximal independent subsets in a greedy manner and tagging the resulting set is the optimal solution to PPM.

Lemma 4.3 *If $t_i \in \mathcal{TS}$ is the first task greedily tagged as POI, which satisfies that $\{t_i\}$ is independent, then there exists a maximal independent set A including t_i, i.e., $t_i \in A$.*

Proof We prove this lemma by construction. Suppose that $B \in \mathcal{I}$ is a maximal independent set, if $t_i \in B$, let $A = B$ and lemma holds. If $t_i \notin B$, let $A = \{t_i\}$. Under this condition, if $|A| = |B|$, then the lemma holds. Otherwise, $|B| > |A|$. We repeatedly and greedily add an element from B into A according to the commutativity of $\mathcal{M}$ until $|A| = |B|$. At this time, there must exist a $y \in B$ and $y \notin A$, satisfying $A = B - \{y\} \cup \{t_i\}$. Tagging the tasks in A results in an objective in PPM, denoting $obj(A)$. Since $obj(t_i)$ is minimum in single independent subset in $\mathcal{I}$, we have $obj(A) \le obj(B)$. On the other hand, B is a maximal independent set, $obj(B) \le obj(A)$. Therefore, $obj(A) = obj(B)$, which means A is also a maximal independent set including t_i.   □

Lemma 4.4 *If there does not exist a task t_i that can be added into ϕ while keeping the independence of the set, then there is no independent set $A \in \mathcal{I}$ including t_i.*

Proof We prove this lemma by contradiction. First of all, we can easily see that $\phi \in \mathcal{I}$ holds by the virtue of transitivity of $\mathcal{M}$. Supposing that $t_i \in A$, $\{t_i\}$ is independent set according to commutativity of $\mathcal{M}$. This violates the condition that t_i cannot be added into ϕ while keeping the independence of the set.   □

Lemma 4.5 *If t_i is the first task greedily selected as POI in $\mathcal{M}$, then $\mathcal{M}$ can be simplified as $\mathcal{M'} = (\mathcal{TS'}, \mathcal{I'})$ where*

$$\begin{aligned}
\mathcal{TS'} &= \{y | y \in \mathcal{TS} \text{ and } \{t_i, y\} \in \mathcal{I}\}, \\
\mathcal{I'} &= \{B | B \subseteq \mathcal{TS} - \{t_i\} \text{ and } B \cup \{t_i\} \in \mathcal{I}\}
\end{aligned} \tag{4.11}$$

Proof If A is a maximal independent set in $\mathcal{M}$, then $A' = A - \{t_i\}$ is independent set in $\mathcal{M'}$ according to second equation in (4.11). Vice versa, any independent set A' in $\mathcal{M'}$ can construct independent set $A = A' \cup \{t_i\}$ in $\mathcal{M}$. Under this circumstance, we have $obj(A) = obj(A') + \alpha$ where $\alpha = N\widetilde{x}_i(1)\rho_i + P_{app}$ is the constant value when tagging t_i. Therefore, a maximal independent set in $\mathcal{M}$ includes the maximal independent set in $\mathcal{M'}$ if it includes t_i.   □

Based on the above lemmas, we design a greedy algorithm for PPM as shown in Algorithm 4.1. Lines 1–4 initialize some variables for greedily constructing maximal tagging set S in lines 5–9. Line 10 calculates the corresponding rewards profile and line 11 returns the results. Under the worst case, the loop runs M times. In the loop, we have to calculate degraded QP using KKT conditions in $O(M \log M)$ according to the Lemma 3 of [33]. Thus, the complexity of Algorithm 4.1 is $O(M^2 \log M)$.

Algorithm 4.1 GreedyPPM

Input: N, M, c_{max}, κ, user set $\mathcal{U}$ and task set $\mathcal{TS}$, POI-tagging price P_{app}, App utility V, network
 effect factor θ and congestion effect factor δ
Output: $p^*(S)$ and $r^*(S)$
1: Let $C - \mathcal{TS}$ be the candidate set
2: Let $S = \phi$ be the tagging set and initialize it to be none.
3: Tag S results in $r(S)$
4: Let $tobj = +\infty$
5: **while** $obj(r(S)) < tobj$ **do**
6:     $tobj = obj(r(S))$
7:     $t_i = \arg\min_{i \in C} obj(r(S \cup \{i\}))$
8:     $S = S \cup \{t_i\}$ and $C = C - \{t_i\}$
9: **end while**
10: solving PPM with $r(S)$ derives $p^*(S)$
11: **return** $p^*(S)$ and $r^*(S)$

Theorem 4.1 *GreedyPPM derives the optimal solution for PPM.*

Proof According to Lemma 4.3, the first task selected by GreedyPPM is included
in the optimal tagging set. Lemma 4.4 indicates that the tasks unselected by
GreedyPPM are not included in the optimal tagging set. Thus, selecting the best
tasks from the remaining tasks is a sub-problem of PPM. Lemma 4.5 guarantees the
correctness of GreedyPPM in the sub-problem of PPM. Therefore, this theorem can
be proved by induction.   □

4.3.3 Stage I: POI-Tagging Pricing

In this section, we analyze how the App makes decisions on P_{app}, which is the
POI-tagging price for helping the platform to tag each task. That is, App selects an
optimal $P_{app}^* = \arg\max \Phi_{app}$ to maximize its payoff defined in Eq. (4.3) under the
anticipation of the results in Stage II and Stage III.

However, the App has a cost of C_{tag} to tag each POI. App's pricing should
guarantee non-negative POI-tagging revenues. Furthermore, we assume that the
platform will cooperate with the App if and only if its POI-tagging revenues are
not the main profit for the App because users are the App's major market instead
of the platform. That is, POI-tagging revenue should be less than advertisement
revenue. Considering these constraints, we mathematically formulate App's POI-
tagging pricing problem as follows:

$$\max \ P_{app}^T r^*(P_{app})$$

$$(APM) \quad s.t. \quad \begin{cases} P_{app} \|r^*(P_{app})\|^2 \leq aN\kappa, \\ P_{app} \geq C_{tag}, \end{cases}$$

where $\|r^*(P_{app})\|^2$ is the square of 2-norm of vector $r^*(P_{app})$.

Directly solving APM is challenging because it requires parameter $r^*(P_{app})$ from Stage II but solving PPM depends on the value of P_{app}. We are inspired by the column generation in the cutting stock problem [34], which iteratively solves the master model and updates the sub-model until they are converged. Therefore, we define PPM and APM as the master model and sub-model. We summarize our approach in Algorithm 4.2:

Algorithm 4.2 APM

Input: N, M, c_{max}, κ, user set $\mathcal{U}$ and task set $\mathcal{TS}$, POI-tagging price P_{app}, App utility V, network effect factor θ and congestion effect factor δ, unit profit of advertisement a, POI-tagging cost C_{tag}

Output: P_{app}

1: assign initial value for $P_{app} \in [C_{tag}, +\infty)$
2: **repeat**
3:    solve PPM through applying Algorithm 4.1 to derive $r^*(P_{app})$ based on P_{app}
4:    solve APM to derive $P_{app} = P^*_{app}$ based on $r^*(P_{app})$
5: **until** the error of best results from two successive iterations satisfies threshold requirement
6: **return** P_{app}

Lemma 4.6 *Algorithm 4.2 will converge to the optimum if and only if the number of iterations is large enough.*

Proof In PPM, the number of tasks to be tagged will decrease monotonously with P_{app}, but increase monotonously in APM. This is, we model their interaction using a non-cooperative Stackelberg game. Therefore, there exists an equilibrium (P^*_{app}, r^*).   □

4.4 Performance Evaluation

In this section, we conduct simulations to evaluate the performance of the proposed PTASIM. The simulation detailed settings are listed in Table 4.2. We run each simulation 100 times and take the average of the results. We implement our formulated model based on IBM ILOG CPLEX Optimization tool (v12.8.0).

Table 4.2 Simulation parameters

Parameter	Value	Parameter	Value
N	200	M	50
c_{max}	80	κ, I_i	$\sim U[0, 1]$
δ	13.14	θ	$\sim U(0, 0.1]$
V	100	C_{tag}	15
a	82.4		

4.4.1 Performance Comparison

To demonstrate the superiority of the proposed greedy approach for PPM in Stage II, we compare our method with a randomized approach (which is widely used in practical applications such as Waze [35]). The key idea of the randomized approach is that each task can be POI by probability and randomly assign each element of r to be 1 (denote as $r^§$). The randomized approach repeats randomizing $r^§$ until $obj(r^§) < \min\{obj(0), obj(1)\}$ holds. We name this method as RandomizedPPM.

4.4.2 Performance Evaluation for Platform

We evaluate computing performance for GreedyPPM (Algorithm 4.1) with comparison to EXPPM and RandomizedPPM. The computing performance is verified from two cases ($M = 10$ and $M = 15$). The total tasks' cost is annotated above the bars. Figure 4.2 shows the computing time of EXPPM is longer than those in the other two cases. As the number of tasks increases, the computing time for EXPPM exponentially increases while the computing time for GreedyPPM and RandomizedPPM still polynomially increases. More importantly, GreedyPPM derives the same tasks' costs as EXPPM, which presents the superiority of Algorithm 4.1. Therefore, we use GreedyPPM as a substitution for EXPPM and conduct the following simulations based on GreedyPPM and use RandomizedPPM for comparison.

We evaluate the platform's strategy from two aspects: how the platform's selection of the number of tasks to be tagged (denote as $||r||^2$) influences the total tasks' cost (including direct recruiting cost and POI-tagging cost) and how

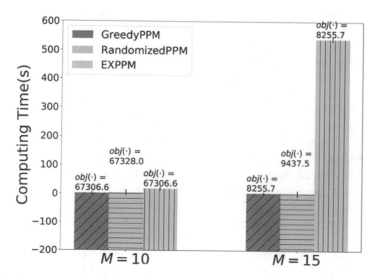

Fig. 4.2 The performance comparison for Algorithm 4.1

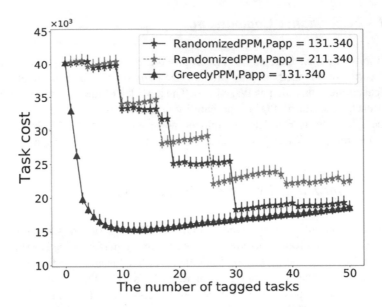

Fig. 4.3 The relationship between the total tasks' cost and the number of tagged tasks

App's intrinsic utility V influences platform's selection on $||r||^2$. The calculation of the total tasks' cost and r is based on the implementation of Algorithm 4.1 and randomized approach. We further compare the performance under cases ($P_{app} = 131.34, \delta = 13.14, P_{app} = 131.34, \delta = 23.14$ and $P_{app} = 231.34, \delta = 13.14$) and range V in [0, 120].

In Fig. 4.3, we can see that there exists an optimal solution for selecting the number of tasks to be tagged. Under the optimal solution, the total tasks' cost is minimized. Selecting all tasks to be tagged or none to be tagged is not necessarily the best choice. The red line with triangle markers represents the policy based on Algorithm 4.1 and the blue and green lines with star markers represent the policies based on the randomized approach. We can see greedy approach (Algorithm 4.1) significantly outperforms the randomized approach in terms of selecting tasks to be tagged as POI in Stage II of the three-stage decision process formulated in the above section. Interestingly, the green dashed line with star markers does not converge to the same point as the other two lines. This is because the POI-tagging price is higher than the other two cases when the platform selects all tasks to be tagged.

In Fig. 4.4, we can observe that more tasks would be selected to be tagged as POI if the App has larger popularity (reflecting in intrinsic utility V). However, if the App increases the POI-tagging price P_{app}, the platform will select fewer tasks to be tagged (indicated in the green line with x markers). Counter-intuitively, the larger congestion effect factor δ has a positive effect on the number of tagged tasks (shown in the red line with triangle markers). This is because the congestion effect affects both App users and normal users. But App users enjoy the intrinsic utility V brought by the App instead of normal users. Therefore, a rational platform will potentially

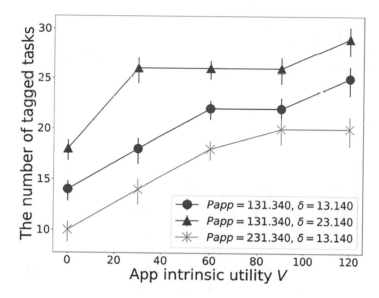

Fig. 4.4 The impact of different parameters on the platform's selection of tasks to be tagged

turn to the App in a situation where the congestion effect is more significant and obvious.

4.4.3 Performance Evaluation for Participation Rate and App

We investigate how incentive reward ρ_i offered by the platform affects the participation rate of both normal users and App users for a given task t_i. Some constant parameters are valued according to Table 4.2 and participation rates are computed based on Lemma 4.1 and 4.2. ρ_i is ranged in [0, 1000] and is sampled 10 times uniformly. The results are shown in Fig. 4.5. The blue line with star markers represents that the participation rate of the task without POI-tagged varies as reward ρ_i while the green and red lines with triangle markers indicate that participation rates of POI-tagged task increase as reward ρ_i. This verifies that incentive ρ_i positively affects users' choice for tasks and hence improves the participation rate of the corresponding task. More importantly, the participation rate of the POI-tagged task is significantly greater than the one without POI-tagging (i.e., $x_i(1) > x_i(0)$). This is because it attracts not only normal users, but also App users to participate in a task once the task is tagged as POI.

We evaluate how App's strategy on POI-pricing P_{app} affects its payoff and the relationship between POI-pricing P_{app} and the number of tagged tasks from the perspective of App and platform (Algorithm 4.2). The payoff calculation is based on Eq. (4.3) and the number of tagged tasks is derived from Algorithm 4.1. Figure 4.6

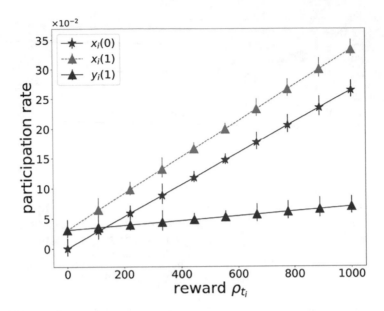

Fig. 4.5 Participation rate under different incentive rewards

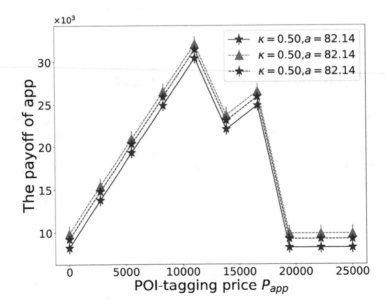

Fig. 4.6 The relationship between App's payoff and its strategy P_{app}

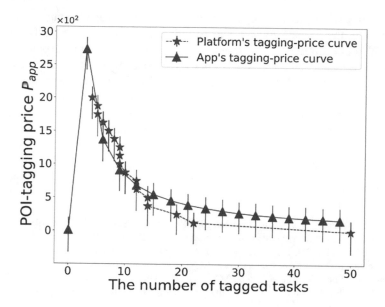

Fig. 4.7 The relationship between App's payoff and its strategy P_{app}

shows the impact of P_{app} on App's payoff under different cases. It can be observed that there exists an optimal P_{app}^* that maximizes App's payoff. This is because larger P_{app} enables the platform to select fewer tasks to be tagged and smaller P_{app} contributes less to the App's total payoff, although the number of tagged tasks increases. In addition, App's payoff benefits from a larger fraction of App users (comparison of the green line with $\kappa = 0.5$ and the blue one with $\kappa = 0.6$). We can also observe that unit advertising profit a has a positive effect on App's payoff (comparison of the blue line with $a = 82.14$ and the red one with $a = 92.14$). Figure 4.7 shows the equilibrium for P_{app} and $||r||^2$. The red solid line represents the App's tagging-price curve while the blue dashed line represents the platform's tagging-price curve. On the one hand, if P_{app} decreases, the platform will select more tasks to be tagged. On the other hand, the App would like to first increase P_{app} for less tagged tasks and then decrease P_{app} for more tagged tasks in order to maximize its payoff. The intersecting point $(12,600)$ is the stable equilibrium.

4.5 Conclusion

In this chapter, we propose PTASIM, an incentive mechanism in MEC that exploits the hybrid platform cooperation for efficient cost reduction and effective participant recruitment. We model the interactions of PTASIM as the three-stage decision process to optimize their strategies and maximize total social warfare. We conduct simulations to evaluate the designed incentive mechanism and the numerical results demonstrate the outperformance and superiority of the proposed scheme. As a potential future direction, it is promising to extend to multiple Apps scenarios. Second, we should take the heterogeneity of the network effect coefficient into account to conduct a general analysis. This will influence the existing results in this work. Third, it is interesting to consider a time-varying scenario which captures practical factors. Last but not least, the practical interaction between the platform and the edge server should be further investigated in the future.

References

1. Wang, J., Wang, Y., Zhang, D., Goncalves, J., Ferreira, D., Visuri, A., Ma, S.: Learning-assisted optimization in mobile crowd sensing: a survey. In: IEEE TII (2018)
2. Chen, H., Li, F., Hei, X., Wang, Y.: Crowdx: Enhancing automatic construction of indoor floorplan with opportunistic encounters. ACM UbiComp 2(4), 159 (2018)
3. Wang, D., Peng, Y., Ma, X., Ding, W., Jiang, H., Chen, F.; and Jiangchuan Liu. Adaptive wireless video streaming based on edge computing: Opportunities and approaches. IEEE Trans. Serv. Comput. 12(5), 685–697 (2018)
4. Zhou, Z., Chen, X., Li, E., Zeng, L., Luo, K., Zhang, J.: Edge intelligence: Paving the last mile of artificial intelligence with edge computing. Proc. IEEE 107(8), 1738–1762 (2019)
5. Marjanović, M., Antonić, A., Žarko, I.P.: Edge computing architecture for mobile crowdsensing. IEEE Access 6, 10662–10674 (2018)
6. Li, T., Qiu, Z., Cao, L., Li, H., Guo, Z., Li, F., Shi, X., Wang, Y.: Participant grouping for privacy preservation in mobile crowdsensing over hierarchical edge clouds. In: IPCCC (2018)
7. Wang, D., Fan, J., Xiao, Z., Jiang, H., Chen, H., Zeng, F., Li, K.: Stop-and-wait: discover aggregation effect based on private car trajectory data. IEEE Trans. Intell. Trans. Syst. 20(10), 3623–3633 (2018)
8. Zhang, X., Yang, Z., Sun, W., Liu, Y., Tang, S., Xing, K., Mao, X.: Incentives for mobile crowd sensing: A survey. IEEE Commun. Surv. Tutorials 18(1), 54–67 (2016)
9. Guo, B, Liu, Y, Wang, L., Li, V.O.K., Jacqueline, C.K., and Yu, Z.: Task allocation in spatial crowdsourcing: current state and future directions. IEEE Internet Things 5, 1749–1764 (2018)
10. Wang, L., Yu, Z., Zhang, D., Guo, B., Liu, C.H.: Heterogeneous multi-task assignment in mobile crowdsensing using spatiotemporal correlation. In: IEEE TMC (2018)
11. Colley, A., Thebault-Spieker, J., Lin, A.Y., Degraen, D., Fischman, B., Häkkilä, J., Kuehl, K., Nisi, V., Nunes, N.J., Wenig, N., et al.: The geography of pokémon go: beneficial and problematic effects on places and movement. In: ACM CHI (2017)
12. Ren, J., Guo, H., Xu, C., Zhang, Y.: Serving at the edge: A scalable iot architecture based on transparent computing. IEEE Netw. 31(5), 96–105 (2017)
13. Zhang, D., Tan, L., Ren, J., Awad, M.K., Zhang, S., Zhang, Y., Wan, P.-J.: Near-optimal and truthful online auction for computation offloading in green edge-computing systems. In: IEEE TMC (2019)
14. Tang, W., Ren, J., Zhang, Y.: Enabling trusted and privacy-preserving healthcare services in social media health networks. IEEE TMM 21(3), 579–590 (2018)
15. Zhu, L., Zhang, C., Xu, C., Sharif, K.: Rtsense: Providing reliable trust-based crowdsensing services in CVCC. IEEE Netw. 32(3), 20–26 (2018)
16. Zhang, C., Zhu, L., Xu, C., Liu, X., Sharif, K.: Reliable and privacy-preserving truth discovery for mobile crowdsensing systems. IEEE TDSC (2019)
17. Yang, D., Xue, G., Fang, X., Tang, J.: Crowdsourcing to smartphones: incentive mechanism design for mobile phone sensing. In: ACM MobiCom (2012)

18. Gao, L., Hou, F., Huang, J.: Providing long-term participation incentive in participatory sensing. In: IEEE INFOCOM (2015)
19. Zhang, X., Yang, Z., Zhou, Z., Cai, H., Chen, L., Li, X.: Free market of crowdsourcing: Incentive mechanism design for mobile sensing. IEEE TPDS 25(12), 3190–3200 (2014)
20. Chen, Y., Li, B., Zhang, Q.: Incentivizing crowdsourcing systems with network effects. In: IEEE INFOCOM (2016)
21. Cheung, M.H., Hou, F., Huang, J.: Make a difference: Diversity-driven social mobile crowdsensing. In: IEEE INFOCOM (2017)
22. Zhang, X., Xue, G., Yu, R., Yang, D., Tang, J.: Truthful incentive mechanisms for crowdsourcing. In: IEEE INFOCOM (2015)
23. Jin, H., Su, L., Nahrstedt, K.: Centurion: incentivizing multi-requester mobile crowd sensing. In: IEEE INFOCOM (2017)
24. Liu, Y., Guo, B., Wang, Y., Wu, W., Yu, Z., Zhang, D.: Taskme: multi-task allocation in mobile crowd sensing. In: ACM UbiComp (2016)
25. Zhang, X., Yang, Z., Liu, Y., Tang, S.: On reliable task assignment for spatial crowdsourcing. IEEE Trans. Emerg. Topics. Comput. 7(1), 174–186 (2016)
26. Yu, H., Iosifidisy, S., Biying, L., Huang, J.: Market your venue with mobile applications: Collaboration of online and offline businesses. In: IEEE INFOCOM (2018)
27. Yu, H., Cheung, M.H., Gao, L., Huang, J.: Economics of public Wi-Fi monetization and advertising. In: IEEE INFOCOM (2016)
28. Gong, X., Duan, L., Chen, X., Zhang, J.: When social network effect meets congestion effect in wireless networks: data usage equilibrium and optimal pricing. IEEE JSAC 35(2), 449–462 (2017)
29. Zhang, M., Gao, L., Huang, J., Honig, M.: Cooperative and competitive operator pricing for mobile crowdsourced internet access. In: IEEE INFOCOM (2017)
30. Nisan, N., Roughgarden, T., Tardos, E., Vazirani, V.V.: Algorithmic Game Theory. Cambridge University Press, Cambridge (2007)
31. Fudenberg, D., Tirole, J.: Game theory. Technical Report, The MIT Press, 1991
32. Lawler, E.L.: Combinatorial Optimization: Networks and Matroids. Courier Corporation, Chelmsford (1976)
33. Yu, H., Neely, M.J.: A new backpressure algorithm for joint rate control and routing with vanishing utility optimality gaps and finite queue lengths. IEEE/ACM ToN 26(4), 1605–1618 (2018)
34. Gilmore, P.C., Gomory, R.E.: A linear programming approach to the cutting-stock problem. Oper. Res. 9(6), 849–859 (1961)
35. Vasserman, S., Michal F., Avinatan H.: Implementing the wisdom of waze. In: Twenty-Fourth International Joint Conference on Artificial Intelligence (IJCAI), pp. 660–666. AAAI Press (2015)

Chapter 5
Coopetition-Aware Incentive Mechanism for Mobile Crowdsensing

Abstract Most of the existing works on MCS only consider designing incentive mechanisms for a single MCS platform. In this chapter, we study the incentive mechanism in MCS with multiple platforms under two scenarios: competitive platform and cooperative platform. We correspondingly propose new competitive and cooperative mechanisms for each scenario. In the competitive platform scenario, platforms decide their prices on rewards to attract more participants, while the users choose which platform to work for. We model such a competitive platform scenario as a two-stage Stackelberg game. In the cooperative platform scenario, platforms cooperate to share sensing data with each other. We model it as many-to-many bargaining. Moreover, we first prove the NP-hardness of exact bargaining and then propose heuristic bargaining. Finally, numerical results show that (1) platforms in the competitive platform scenario can guarantee their payoff by optimally pricing on rewards and participants can select the best platform to contribute; (2) platforms in the cooperative platform scenario can further improve their payoff by bargaining with other platforms for cooperatively sharing collected sensing data.

Keywords Multiple platforms · Coexistence · Competitive interaction · Cooperative interaction · Two-stage Stackelberg game · Many-to-many bargaining

5.1 Introduction

5.1.1 Motivations and Challenges

With the advent and pervasiveness of mobile devices, mobile crowdsensing (MCS) [1, 2], as a new sensing paradigm, has increasingly attracted much attention from the research community. It makes full use of a crowd of mobile devices (equipped with rich sensors, deployed on-site and held by people) to conduct sensing tasks and collect sensing data. There have been many applications of mobile crowdsensing, such as Waze [3] for real-time traffic monitoring, Gigwalk [4] for mobile market research, U-Air [5] for air quality monitoring, and FlierMeet [6] for public information sharing.

The incentive mechanism is one of the major challenges in mobile crowdsensing because users sacrifice some efforts (computing and communicating energy) in general crowdsensing and especially suffer from privacy concerns in location-aware crowdsensing. Users do not voluntarily participate in crowdsensing unless they are motivated enough to compensate their sensing cost [7]. A lot of works [8–16] are devoted to studying and designing appropriate and efficient incentive mechanisms [17] to reward participants and further attract more users for mobile crowdsensing. However, most of them focus on incentive mechanisms for a single platform and assume the platform to be a central controller which coordinates the data and service exchanges between users and sensing service requesters. With the increasing development of mobile crowdsensing, data sources from mobile devices become heterogeneous and service requests from subscribers are getting diverse. It imposes a potential challenge to the platform to bridge the gap between mobile users and service subscribers. On the other hand, it probably gives rise to overloading the central platform. Therefore, it is imperative to develop multiple MCS platforms to cope with these new challenges.

5.1.2 Contributions

In this chapter, we will study how to design incentive mechanisms for the MCS system with multiple platforms, where there exist many heterogenous platforms recruiting limited participants for different sensing tasks. We divide the situation into *competitive platform scenario* and *cooperative platform scenario*. The former one focuses on the interaction among mobile participants and platforms for data collection, which addresses how the platforms price rewards and how the participants select the best platform to work for. The latter one pays attention to the interaction among platforms for sharing the common and reusable sensing data, which addresses how the platforms cooperate with each other and at what prices.

For the competitive platform scenario, our proposed incentive mechanism focuses on the interactions between platforms and participants. Both of them make decisions to maximize their own payoffs in mobile crowdsensing. For platforms, they subtly price on rewards to recruit enough participants for their task execution. Participants, intentionally select the best platform to exert efforts for more revenues. Unfortunately, the competition that arose from user recruitment in multiple platforms complicates the design of the incentive mechanism. In this chapter, we use a two-stage Stackelberg game to model the competition. Nash equilibrium indicates how platforms and users can optimally make decisions under a multiple platform environment. Obviously, the competition reduces the revenues platforms received, hence imposing a negative externality to the competitive mechanism among platforms. This negative externality calls for a new mechanism to guarantee and improve platforms' payoffs under multiple MCS platforms.

Therefore, for the cooperative platform scenario, we design a new incentive mechanism that focuses on the interactions among platforms after task completion. We allow platforms to cooperate with each other by sharing the sensing data collected from participants. That means one platform can buy sensing data from other platforms. In order to maximize the payoff in the cooperative scenario, each platform subtly determines which platform to cooperate with and at what price. However, the detailed economic interaction among platforms for cooperation has not been sufficiently explored. In this chapter, we use Nash bargaining theory to model the cooperative platform scenario. Moreover, we extend one-to-one Nash bargaining to one-to-many bargaining, and further many-to-many bargaining for multiple platform cooperation. Through deriving NBS (Nash Bargaining Solution), our proposed mechanism addresses how a platform cooperates, i.e., which one to cooperate with and how much to pay for cooperation.

In summary, we consider the incentive design for the MCS system with competitive and cooperative platforms. The main contributions of the proposed competitive and cooperative mechanisms for multiple crowdsensing platforms are summarized as follows:

- *Novel incentive mechanisms for multiple crowdsensing platforms*: Different from traditional single central platform MCS, we propose an incentive mechanism that works with multiple crowdsensing platforms. The proposed incentive mechanisms not only address the interactions between platforms and participants, but also handle the cooperation among platforms.
- *Modeling and theoretical analysis for the competitive and cooperative mechanisms*: We systematically model the scenarios for multiple crowdsensing platforms and design corresponding incentive mechanisms, called **MP-Coopetition**, for both competitive and cooperative scenarios. We analyze the competitive mechanism and the cooperative mechanism by leveraging the two-stage Stackelberg game and Nash Bargaining, respectively.
- *Extensive simulations for performance evaluations*: We conduct simulations for the impact of multiple platforms on mobile crowdsensing over traces and the performance of the proposed competitive and cooperative mechanisms. The results show the outperformance of our proposed incentive mechanisms in both competitive and cooperative scenarios.

5.1.3 Related Works

This chapter focuses on designing effective and efficient incentive mechanisms for MCS systems. We now briefly review some related works in both MCS and other related areas.

Single Platform Incentives Incentive mechanism design is one of the major challenges and research focuses of MCS research [1, 7, 14, 15]. Several existing works of literature [8–12, 18–23] study incentive mechanism in different ways and under different scenarios. Such as, [8] investigates the Stackelberg game and auction for the platform-centric scenario and user-centric scenario, respectively. Wen et al. [21], Zhang et al. [22], Wang et al. [18, 23], Hu et al. [9], Gao et al. [10], Jin et al. [12], and Duan et al. [19] also study auction-based incentive mechanisms with different economic properties. Chen et al. [11] studies incentive mechanism with network effect. All of these works assume that a central platform uses incentive mechanisms to coordinate other players. Different from them, we focus on designing an effective incentive mechanism working for multiple crowdsensing platforms. Multiple platforms make incentive mechanism design much more challenging and complicated to analyze.

Multi-Tasks and Multi-Requesters Many recent MCS systems [12, 24–26] also consider to support multi-requesters and multi-tasks from different platforms or systems, but most of them do not consider multiple sensing platforms. Jin et al. [12] considers MCS systems with multiple data requesters and proposes an incentive mechanism based on double auction which is integrated with data aggregation over workers' data. Liu et al. [24] considers two typical multi-task allocation situations in MCS with bi-objective optimization goals (such as minimizing movement distance while maximizing accomplished tasks or minimizing incentive payments). Two task allocation algorithms are proposed for each type of situation. Li et al. [25] also studies participant selection with multiple heterogeneous sensing tasks and proposes to reuse cached sensing data to fulfill more tasks with fewer selected participants. Jarrett et al. [26] addresses how to recommend multiple tasks to workers using recommendation technique in a distributed manner. However, most of these works do not study how to incentivize users to perform tasks.

Competition and Cooperation There are some existing literatures studying competition and cooperation [22, 27–31] for different systems. Yu et al. [27] and Gao et al. [29] exploit the cooperation using a one-to-many bargaining game for Wi-Fi deployment and mobile data offloading, respectively. Zhang et al. [28] introduces a pricing framework for different mobile network operators to price on traffic under competitive scenario and cooperative scenario using a two-stage game while we use two-stage Stackelberg game (non-cooperative game) for competitive platform scenario and Nash bargaining game (cooperative game) for cooperative platform scenario. Yu et al. [30] proposes a novel spectrum sharing framework for cooperation and competition between LTE and Wi-Fi in the unlicensed band based on second-price reverse auction to enables the LTE provider and the Wi-Fi Access Point Owners (APOs) to effectively negotiate the operation mode (cooperation and competition). However, its main work assumes that only one LTE provider bids with many APOs. Although it extends to handle multiple LTE providers' scenarios by running an auction for each LTE provider, the scenarios and technical details it considers are different from ours, thus its method cannot be directly applied to MCS with multiple platforms. Li et al. [31] studies an incentive mechanism to handle the competition among users with close geographical positions, which results in economic waste. Truthful incentive mechanism in [22] investigates the competition and cooperation among service providers (crowd workers) instead of sensing platforms. Different from them, our work (1) focuses on the incentive mechanism design in MCS with multiple platforms; (2) comprehensively studies competitive and cooperative mechanisms for incentive; (3) extends one-to-many bargaining to many-to-many bargaining for platforms to cooperate; and (4) proves the NP-hardness of sequential bargaining and proposes corresponding heuristic bargaining. Dustdar et al. [32], Dustdar and Truong [33], and Hoenisch et al. [34] also consider the cooperation between humans and machines in cloud computing by integrating human-based computing elements into machine-based computing elements via elastic processes. These works focus on how to virtualize human-based computing and process complex computing tasks requiring human intelligence. The cooperations are based on voluntary workers, which is different from our setting. In addition, many crowdsensing tasks usually do not require human intelligence.

Data Trading Although data trading [35, 36] is similar to our cooperative mechanism for data sharing among multiple crowdsensing platforms, their methodology and scenarios are different from ours. Yu et al. [35] studies the brokerage-based market for the trading platform to match the data market supply and demand between sellers and buyers, which further utilizes prospect theory and expected utility theory to model the users' realistic trading behaviors. Ma et al. [36] optimally designs time and location-aware mobile data pricing to incentivize users to smooth traffic in peak hours and hot spots and reduce network congestion. They both use a two-stage decision process to formulate their problems and study data pricing while our problem in the cooperative platform scenario is modeled as many-to-many bargaining. Besides, they also assume a central platform to price the shared data. Therefore, their research techniques can not directly be applied to our problem in the cooperative platform scenario.

Incentive Type and Implementation In this chapter, we consider using payments to provide monetary incentives for participation. The payments can also be transformed into credits or other virtual rewards. But voluntarism as a self-motivated incentive can also be used for mobile crowdsensing, such as collecting emergent information in natural disasters scenario [37]. Scekic et al. [38] surveys different incentive mechanisms widely used in social computing, including non-cash incentives like psychological voluntarism and altruism. For implementation, the existing programmable frame (such as PRINC [39]) or collaborative crowdsourcing infrastructure (such as [40]) can be utilized to realize our proposed incentive mechanism.

5.2   Game Modeling

5.2.1   System Overview

In general, a mobile crowdsensing system consists of requesters, a platform, and mobile users. In this chapter, we consider an MCS system with multiple platforms, where each platform can publish sensing tasks and recruit users to complete them. Since requesters only purchase services from platforms, we only study strategies and interactions among platforms and users. More specifically, we assume there is a set $\mathcal{I} = \{1, 2, \ldots, I\}$ of users, each of whom can perform one or several sensing tasks. Besides, there are multiple platforms residing in the cloud-based MCS system denoting as $\mathcal{K} = \{1, 2, \ldots, K\}(K \geq 2)$. Mobile users are distributed in a variety of points of interest (PoI) while each platform recruits a group of users in specific PoI to perform sensing tasks and collect sensing data. We assume each user i can only be selected by a platform and leave cross-platform crowdsensing for future work.

When user i is recruited by platform j, it contributes effort level $x_{j \leftarrow i}$ and receives payment r_j. Due to limited energy, $x_{j \leftarrow i}$ has a bound B_i^U ($\forall i \in \mathcal{I}, \forall j \in \mathcal{K}$). At the same time, user i will incur some cost for performing sensing tasks from platform j, which is denoted as $c_{j \leftarrow i}$. Here, $c_{j \leftarrow i}$ may be various over different users and platforms, and become their private information. Platforms can estimate it through history. Each user works for a selected platform and each platform may have several users selected for their sensing tasks. Let I^j be the worker set for platform j and $\sigma(i)$ be the platform user i works for.

The general interactions among multiple platforms and users are illustrated in Fig. 5.1. At the beginning, platforms announce rewards to recruit users to participate in their own sensing tasks meanwhile describing the requirements of sensing tasks (such as how much data to collect, how long to perform a task, and how much effort to exert). Facing the rewards and sensing tasks from platforms, users select the best platform to contribute to sensing efforts for more revenues. After collecting sensing data, platforms bargain with each other to share sensing data because this cooperation will further compensate for the loss of competition during recruiting participants and improve their payoffs.

We consider a delay budget T_d for all sensing tasks due to the time-sensitive property. Furthermore, we assume the system model of crowdsensing is quasi-static, which means that the system parameters (e.g., unit cost or value) and settings (e.g., delay budget or the number of users and platforms) remain almost the same for a period. We summarize the key notations in this chapter in Table 5.1.

5.2.2   Crowd Workers

Crowdsensing enables users to contribute their sensing capability, hence being crowd workers. For ease of exposition, we use "user", "participant" and "crowd

Fig. 5.1 MP-Coopetition architecture with multiple crowdsensing platforms

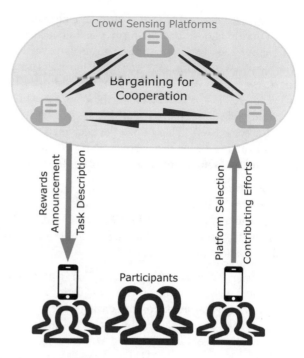

Table 5.1 Notations used in Chap. 5

$i, I, \mathcal{I}$	User index, number of users and users set
$j, K, \mathcal{K}$	Platform index, no. of platforms and platforms set
B_i^U	User i's efforts bound
$x_{j \leftarrow i}$	The efforts user i exerts to platform j
$c_{j \leftarrow i}$	The sensing cost user i suffered in platform j
r_j	The rewards offered by platform j
B_j^P	Platform j's rewards budget
Ψ_i^u	User i's payoff
I^j	Platform j's participants set
Φ_j	Platform j's SoV
Ψ_j^{P-}	Platform j's payoff under competition
$\mathcal{B}^j$	The bargainer set for j with size of N^j
b_k^j	Indicator of the cooperation b/w platform j and k
$p_{k \leftarrow j}$	Payment platform j give to k if cooperating
$X_{j \leftarrow k}$	Amount of data k hold and bargain with j
E_k	Operational and commu. cost k suffered if cooperated
Ψ_j^{P+}	Platform j's payoff under cooperation
V_k^j	Platform k's payoff when bargaining with j

worker" interchangeably in this chapter. Under the scenario of multiple platforms, user i has to choose one platform j and contribute to it with rewards r_j. Due to rationality, each user aims to maximize its payoff by choosing the platform with the most reward. This may cause all the users to choose the platform with the most reward. However, there is working cost $c_{j \leftarrow i}$ for user i working in platform j ($\forall i \in \mathcal{I}, \forall j \in \mathcal{K}$). Therefore, the platform with the most reward may not be necessarily the best choice for all users, as this platform may incur a more working cost for some users. This brings the challenge, which is to choose an appropriate platform to maximize the users' payoff. When user i exerts $x_{j \leftarrow i}$ to platform j, there are two factors affecting its payoff as follows:

Rewards Under Effort Level $x_{j \leftarrow i}$, $R_i^U(x_{j \leftarrow i})$ Based on the payment r_j offered by platform j, if user i exerts effort level $x_{j \leftarrow i}$ to platform j, then it receives a total reward $r_j x_{j \leftarrow i}$, denoted by $R_i^U(x_{j \leftarrow i})$. Rewards offered by the platform will stimulate the user to contribute and increase its utility. Therefore, this factor will positively affect the user's payoff.

Cost Under Effort Level $x_{j \leftarrow i}$, $C(x_{j \leftarrow i})$ Due to the working cost, when user i contributes $x_{j \leftarrow i}$ amount of effort to platform j, it suffers a total cost (e.g., the energy or time consumed for the sensing tasks from the platform) in proportional to its effort level $x_{j \leftarrow i}$, denoted by $C(x_{j \leftarrow i})$. In this chapter, we consider the convex cost scenario where marginal cost is increasing with effort level $x_{j \leftarrow i}$. We define total cost as $c_{j \leftarrow i} x_{j \leftarrow i}^2$ [41], i.e., $C(x_{j \leftarrow i}) = c_{j \leftarrow i} x_{j \leftarrow i}^2$. Cost $C(x_{j \leftarrow i})$ will do harm to the user's enthusiasm and decrease its payoff. Therefore, this factor will negatively affect the user's payoff.

Overall, we define the payoff of user i under effort level $x_{j \leftarrow i}$ and rewards r_j when it works in platform j as the utility gain it receives. i.e.,

$$\Psi_i^u(r_j, x_{j \leftarrow i}) = R_i^U(x_{j \leftarrow i}) - C(x_{j \leftarrow i}). \tag{5.1}$$

5.2.3   *Platforms and Negative Externalities*

Compared to the conventional single platform of mobile crowdsensing, we study multiple platforms mechanism in this chapter. Under a certain population of users, each platform has to recruit as many users to be participants in performing sensing tasks as possible. Therefore, multiple platforms impose a competitive relationship. Each of them wants to recruit more participants for a higher payoff and takes the best action by announcing optimal rewards r_j. Note that this kind of competition in recruiting participants is a negative externality. Therefore, our analysis of maximizing payoff is getting more challenging.

5.2.3.1 Competitive Platform Scenario

Under a competitive platform scenario, each platform j independently decides on pricing rewards r_j for attracting participants in order to maximize its payoff. Let $-j$ denote the platforms other than j and r_{-j} denote the rewards offered by platforms $-j$. If r_j is less than r_{-j}, platform j may get less participants than $-j$; or otherwise.

However, platforms must consider their own rewards budget when pricing rewards r_j. Here we denote B_j^P as the rewards budget for each platform j ($\forall j \in \mathcal{K}$). Supposing that platform j announces its rewards by r_j, there are some users who participate in platform j with $x_{j \leftarrow i} > 0$, denoted by $\mathcal{I}_j$. On the other hand, each platform lacks information about platforms $-j$ and the working cost for all users. From this perspective, it is challenging to price rewards r_j. Thus, it is very important to consider the subtle payoff for platform j. When platform j prices r_j to recruit participants, there are two factors affecting its payoff as follows:

Total Offered Rewards If user i exerts $x_{j \leftarrow i} > 0$ to platform j which contributes to perform its sensing tasks, platform j will reward each user $i \in I^j$ with r_j which is determined by j. This significantly costs the platform if there is no compensation. Therefore, total offered rewards negatively affect platforms' payoff. Formally, we denote the total offered rewards by $\sum_{i \in I^j} r_j x_{j \leftarrow i}$.

Service of Value (SoV) Due to the negative effect of offered rewards, platform j needs to well aggregate the sensing results from participants I^j and sells data-based service to requesters. This can be regarded as compensation for platforms' cost and is formally defined below:

Definition 5.1 (**Service of Value**): The value of sensing results from I^j, which is perceived by platform j, can be characterized by:

$$\Phi_j(x^j) \triangleq \lambda_j \sum_{i \in I^j} x_{j \leftarrow i}, \tag{5.2}$$

where x^j is total effort level $\sum x_{j \leftarrow i}$ from I^j and λ_j is the coefficient for platform j which characterizes the data utility brought by I^j [8, 11].

Overall, we define the payoff of platform j under total effort level x^j from the participants recruited and its pricing rewards r_j when it has recruited participants I^j as the utility gain it receives. i.e.,

$$\Psi_j^{P-}(r_j, x^j) = \Phi_j(x^j) - R_j^{P-}(x^j), \tag{5.3}$$

where $R_j^{P-}(x^j)$ is the total offered rewards under total effort level x^j, i.e., $R_j^{P-}(x^j) = \sum_{i \in I^j} r_j x_{j \leftarrow i}$. Here, the superscript P_- represents the competitive platform scenario and P_+ represents the cooperative platform scenario.

5.2.3.2   Negative Externalities Over Multiple Platforms

Due to the quasi-static property, the number of users in mobile crowdsensing remains constant for some period. That means participants for sensing tasks are limited if multiple platforms simultaneously recruit participants within overlapping delay budget T_d. On the other hand, participants are not allowed to perform sensing tasks across multiple platforms. Therefore, there exists competition among platforms when pricing on rewards to recruit participants as many as possible. And this kind of competition in recruiting participants, furthermore, imposes negative externalities among platforms since each platform's recruitment of more participants by optimal rewarding will significantly decrease other platforms' payoff.

The pricing on rewards with negative externalities substantially challenges the problems studied in this chapter. This challenge drives us to explore a cooperation mechanism among platforms in order to improve the total payoff.

5.2.3.3   Cooperative Platform Scenario

Negative externalities in the competitive platform scenario result in inefficiency for mobile crowdsensing with multiple platforms. On the other hand, each platform has incentives to make full use of sensing results by cooperating with other platforms, since one platform benefits from shared sensing results while another platform receives payments for sharing sensing results, hence avoiding wasting sensing results. Based on this observation, we turn to cooperative game theory and apply Nash Bargaining to address inter-platform cooperation economic issues.

Under the cooperative platform scenario, each platform j coordinates (negotiates) with platform $k \in \mathcal{K}(j \neq k)$ whether to share sensing results in order to further improve their own payoffs. Furthermore, if k agrees with j on sharing sensing results with it, j should give the number of payments to k for cooperative compensation. Based on these interactions, we exploit the Nash Bargaining framework, especially one-to-many bargaining. Due to the analysis of all platforms, we have to conduct once one-to-many bargaining for each platform in order to determine the bargaining outcome.

Before analysis, it is important to define the cooperative payoffs for each platform and analyze it's contributing factors to payoffs. Let $\mathcal{B}^j$ be the bargainers set for j with size of N^j. For $k \in \mathcal{B}^j$, b_k^j is bargaining outcome which indicates if k cooperates with j ($b_k^j = 1$ represents "yes" and $b_k^j = 0$ otherwise), and $p_{k \leftarrow j}$ is the payments for k if $b_k^j = 1$. Note that $p_{k \leftarrow j} = 0$ if j fails to cooperate with k, i.e. $b_k^j = 0$. When j bargains with k, k determines b_k^j and j decides on $p_{k \leftarrow j}$, hence $(b_k^j, p_{k \leftarrow j})$ is bargaining outcome. Moreover, when k decides to cooperate with j, it shares $X_{j \leftarrow k}$ amount of sensing results with j as well as suffers from operational and communication cost E_k. To facilitate the definition of payoffs for cooperative platforms, let $\boldsymbol{b}_{\mathcal{B}^j} \triangleq (b_1^j, b_2^j, \ldots, b_{N^j}^j)$ and $\boldsymbol{p}_{\mathcal{B}^j} \triangleq (p_{1 \leftarrow j}, p_{2 \leftarrow j}, \ldots, p_{N^j \leftarrow j})$ be the bargaining outcome between platform j and all bargainers $\mathcal{B}^j$. It is easy to see that the total shared sensing results $\sum\limits_{k\in\mathcal{B}^j} X_{j\leftarrow k}$ (called benefits) and total offered payments $\sum\limits_{k\in\mathcal{B}^j} p_{k\leftarrow j}$ (called costs) significantly affect the payoff of platform j. On the other hand, the total cost based on operational and communication cost E_k (called costs) and received payments $p_{k\leftarrow j}$ (called benefits) significantly affect the payoff of bargainer k. Therefore, the payoff of platform j is

$$\Psi_j^{P+}(\boldsymbol{b}_{\mathcal{B}^j}, \boldsymbol{p}_{\mathcal{B}^j}) = f\left(\sum_{k\in\mathcal{B}^j} b_k^j X_{j\leftarrow k}\right) - \sum_{k\in\mathcal{B}^j} p_{k\leftarrow j}. \tag{5.4}$$

The payoff of bargainer $k \in \mathcal{B}^j$ is

$$V_k^j(b_k^j, p_{k\leftarrow j}) = p_{k\leftarrow j} - b_k^j E_k X_{j\leftarrow k}. \tag{5.5}$$

Here, we choose an increasing and strictly concave but non-negative function for $f(\cdot)$ (i.e., $f'(z) > 0$, and $f''(z) < 0$) in order to characterize the effect of shared sensing results from bargainers set. Obviously, $f(0) = 0$ holds.

5.2.4 Problem Definition

The problems studied in this chapter are extracted as follows:

Definition 5.2 (**CompPricing**) Given users set $\mathcal{I}$, the aim of **CompPricing** is to address how platforms price the amount of rewards to attract as many users to be participants as possible as well as how each user chooses the best platform to contribute proper sensing efforts for more revenues.

Definition 5.3 (**CoopPricing**) Given a set $\mathcal{B}^j$ of platforms, the aim of **CoopPricing** is to address which platform $k \in \mathcal{B}^j$ platform j should cooperate with in order to receive the shared sensing results and how much platform j should pay to these cooperative platforms.

Lemma 5.1 *CoopPricing is NP-hard.*

Proof To prove CoopPricing is NP-hard, which its formulation is Eq. (5.18), we have to first prove that its decision problem is NP-complete. The decision problem is whether there exists a subset of $\mathcal{B}^j$ cooperating with platform j such that the number of bargainers in the subset is no more than t. For its decision problem, we need to check whether the decision problem belongs to NP or not, then use a well-known NP-complete problem to reduce the decision problem of CoopPricing in polynomial time. Actually, the decision problem of CoopPricing is an instance of the Subset Sum problem, which is a typical NP-complete problem [42]. The decision problem belongs to NP as verifying its solution can terminate in polynomial time. The reduction proceeds as follows: For the bargaining of platform j with bargainers $\mathcal{B}^j$, it is equivalent to selecting a subset from $\mathcal{B}^j$ to cooperate with. Under given bargaining sequence $L^j[1 : N^j]$, bargainers set $\mathcal{B}^j$ can be represented as $\mathcal{SN}_1 = \{(X_{l_1^j}, E_{l_1^j}), (X_{l_2^j}, E_{l_2^j}), \ldots, (X_{l_s^j}, E_{l_s^j}), \ldots, (X_{l_{Nj}^j}, E_{l_{Nj}^j})\}$. The goal of sequential bargaining is that determining who cooperate with and in what price for each $l_s^j \in \mathcal{B}^j$ (i.e., $(b_{l_s^j}^{*j}, p_{l_s^j \leftarrow j}^*)$). Therefore, we calculate $p_{l_s^j \leftarrow j}^*$ and $\Omega_s (\forall l_s^j \in \mathcal{B}^j)$ using Proposition 5.4. This equivalently transforms $\mathcal{SN}_1$ to $\mathcal{SN}_2 = \{p_{l_1^j \leftarrow j}^*, p_{l_2^j \leftarrow j}^*, \ldots, p_{l_s^j \leftarrow j}^*, \ldots, p_{l_{Nj}^j \leftarrow j}^*\}$. Recall the goal of sequential bargaining, we can clearly see that some bargainer $l_s^j \in \mathcal{B}^j$ should be selected from $\mathcal{SN}_2$ such that the condition $p_{l_s^j \leftarrow j}^* \leq min\{r_j^*, \Omega_s\}$ holds. These selected bargainers consist of a subset $\mathcal{SN}_3$ of $\mathcal{SN}_2$ while constraint $\sum_{l_s^j \in \mathcal{SN}_3} p_{l_s^j \leftarrow j}^* \leq min\{|\mathcal{SN}_3|r_j^*, \sum_{l_s^j \in \mathcal{SN}_3} \Omega_s\}$ holds. $\square$

5.3 Detailed Design with Equilibrium Analysis: Competition Among Platforms

In this section, we model the competitive platform scenario using Stackelberg game theory as a two-stage game and further analyze its Nash Equilibrium by applying backward induction.

5.3.1 User's Decision

In this subsection, given the announced rewards from platforms, we first study how user strategizes under single platform j and then extend the analysis to the multiple platform scenario. This is the second stage of the overall Stackelberg game.

5.3.1.1 Strategizing in Single Platform

Given the rewards r_j announced by platform j, each user i strategizes the proper amount of sensing efforts $x_{j \leftarrow i}$ to platform j in order to maximize its payoff (indicated in Eq. (5.1)). At the same time, it should consider its sensing restrict B_i^U. The research concern here is included in **CompPricing** and reformulated mathematically as follows:

$$\textbf{(UOPT)} \quad max \ \Psi_i^u(r_j, x_{j \leftarrow i})$$
$$\text{s.t.} \quad 0 \leq x_{j \leftarrow i} \leq B_i^U, \tag{5.6}$$

where B_i^U is the sensing effort bound for user i, and variable $x_{j \leftarrow i}$ represents the strategy which is the efforts level user i exerting to platform j. Constraint (5.6) demonstrates that user i can not exert more sensing efforts than B_i^U which characterizes the capacity of each user (such as remaining battery level). User i determines the best sensing efforts level $x_{j \leftarrow i}^*$ to platform j by solving **UOPT**.

In this chapter, we assume the convex cost for users in MCS. Hence, the **UOPT** is a convex optimization problem in $x_{j \leftarrow i}$. This is helpful for the analysis of strategizing sensing efforts for users in the second stage based on the rewards announced by platforms in the first stage.

In terms of convex optimization problem, the optimum is dominated by KKT conditions, which is the sufficient and necessary condition for solving convex optimization problem with inequality constraints [43]. On the other hand, solved KKT conditions provide closed-form solution which is helpful to implement the proposed mechanism in computer. By expanding **UOPT**, we can easily derive its KKT conditions, which are given as follows:

$$
\begin{aligned}
r_j - 2c_{j \leftarrow i} x_{j \leftarrow i} + \mu_1^i - \mu_2^i &= 0, \forall i, j, \\
\mu_1^i (x_{j \leftarrow i} - B_i^U) &= 0, \forall i, j, \\
-\mu_2^i x_{j \leftarrow i} &= 0, \forall i, j, \\
\mu_1^i &\geq 0, \forall i, \\
\mu_2^i &\geq 0, \forall i,
\end{aligned} \tag{5.7}
$$

where, μ_1^i and μ_2^i are KKT multiplier.

Based on these KKT conditions, we have the following proposition:

Proposition 5.1 *Based on the rewards r_j announced by platform j, user i strategizes its optimal sensing efforts $x_{j \leftarrow i}^*$ by*

$$
x_{j \leftarrow i}^* = min\{B_i^U, \frac{r_j}{2c_{j \leftarrow i}}\} \tag{5.8}
$$

Proof There are three unknowns in the Eq. (5.7) from KKT conditions. Solving these equations provides a solution for **UOPT**. Hence, this proposition obviously holds.   □

5.3.1.2 Strategizing with Multiple Platforms

Suppose that there are K platforms announcing rewards profile $\boldsymbol{r} = (r_1, r_2, \ldots, r_K)$ to recruit participants, each user has to subtly react to the rewards profile and optimally choose the best platform to contribute within its own sensing capacity B_i^U. We consider user i as an example. Proposition 5.1 points out the best efforts for the maximal payoff if a user chooses a certain platform. Generally speaking, the user will select a platform with high rewards and low sensing costs. When user i determines which platform it should choose to contribute, it simply checks the received payoff under announced r one by one and selects the best platform it desires. It looks like that it independently strategizes its best sensing efforts with each platform according to Proposition 5.1. Therefore, we have the obvious proposition as follows with omitted proof:

Proposition 5.2 (Platform Selection Policy) *The platform j^* to which user i would like to contribute its sensing efforts to is*

$$j^* = \arg\max_{j \in \mathcal{K}} x^*_{j \leftarrow i} \tag{5.9}$$

The detailed method for users to select the best platform is given in Algorithm 5.1.

Algorithm 5.1 Best Platform Selection (BPS)

Input: users set $\mathcal{I}$, platforms set $\mathcal{K}$, rewards profile $r = (r_1, r_2, \ldots, r_K)$, working costs matrix $[c_{j \leftarrow i}]_{|\mathcal{I}| \times |\mathcal{K}|}$, sensing capacity B_i^U, $\forall i \in \mathcal{I}$
Output: $(j_1^*, j_2^*, \ldots, j_{|\mathcal{I}|}^*)$
1: **for** $i \in \mathcal{I}$ **do**
2:     Let $tempX = 0$ and $tempI = 0$
3:     **for** $j = 1$ to $|\mathcal{K}|$ **do**
4:         **if** $\frac{r_j}{2c_{j \leftarrow i}} \geq B_i^U$ **then**
5:             continue
6:         **else if** $tempX < \frac{r_j}{2c_{j \leftarrow i}}$ **then**
7:             $tempX \leftarrow \frac{r_j}{2c_{j \leftarrow i}}$
8:             $tempI \leftarrow j$
9:         **end if**
10:    **end for**
11:    $j_i^* \leftarrow tempI$
12: **end for**
13: **return** $(j_1^*, j_2^*, \ldots, j_{|\mathcal{I}|}^*)$

In Algorithm 5.1, the outer loop traverses user set, while the inner loop derives the best platform for each user according to Proposition 5.2. Lines 4–5 filter the unmeet platform due to higher required sensing efforts and lines 6–8 record the best platform and its index. It is easy to see that the complexity of BPS is $O(|\mathcal{I}| \times |\mathcal{K}|)$.

5.3.2 Platforms' Competitive Pricing

Under the competitive platform scenario, each platform independently determines to price on rewards to recruit participants. With the prediction of the outcome in the second stage of formulated Stackelberg game (Sect. 5.3.1), we study the pricing policy for each platform in the first stage.

The concern included in **CompPricing** is that platform j prices optimal rewards value r_j so that more participants accept these rewards and contribute their sensing efforts to this platform, considering the sensing cost of all users. By optimally pricing, each platform aims to maximize its payoff. Here we mathematically formulate the payoff maximization for each platform. Note that payoff maximization studied here is with respect to the competitive platform scenario. Based on Eq. (5.3), platforms' payoff maximization is formulated as follows:

$$\textbf{(CompOPT1)} \quad \max \ \Psi_j^{P-}(r_j, \boldsymbol{x}^j)$$
$$\text{s.t.} \quad 0 \leq r_j \leq B_j^P, \tag{5.10}$$

where B_j^P is the rewards budget for each platform j, and variable r_j represents the strategy which is the rewards optimally priced to recruit participants. $\boldsymbol{x}^j$ is sensing efforts profile from $\forall i \in \boldsymbol{I}^j$. Constraint (5.10) demonstrates that platform j can not price more rewards than B_j^P which characterizes the budget for each platform j ($\forall j \in \mathcal{K}$). Platform j optimally prices on rewards r_j^* to recruit participants by solving **CompOPT1**.

Solving **CompOPT1** is nontrivial since (1) payoff $\Psi_j^{P-}(r_j, \boldsymbol{x}^j)$ of platform j is not convex in r_j; (2) the participants set $\boldsymbol{I}^j$ is not prior to platform j. This brings the challenge to maximize platforms' payoff by optimally pricing on rewards r_j under the competitive platform scenario.

Fortunately, each platform j can predict $x_{j \leftarrow i} \in \boldsymbol{x}^j$ using the conclusion that sensing efforts $x_{j \leftarrow i}$ is related to r_j and cost $c_{j \leftarrow i}$ (i.e., $x_{j \leftarrow i}^* = \frac{r_j}{2c_{j \leftarrow i}}$) in Sect. 5.3.1.1. On the other hand, platform j will not differentiate participants in $\boldsymbol{I}^j$ and announce the same rewards r_j to all participants. Therefore, platform j only considers specific participant i when pricing on r_j. Substituting $x_{j \leftarrow i}$ with $\frac{r_j}{2c_{j \leftarrow i}}$, **CompOPT1** is transformed into:

$$\textbf{(CompOPT2)} \quad \max \ \lambda_j \sum_{i \in I^j} \frac{r_j}{2c_{j \leftarrow i}} - \sum_{i \in I^j} \frac{r_j^2}{2c_{j \leftarrow i}}$$
$$\text{s.t.} \quad 0 \leq r_j \leq B_j^P. \tag{5.11}$$

CompOPT2 is convex in r_j and its KKT conditions are given as follows:

$$\frac{\lambda_j}{\sum\limits_{i \in I^j} 2c_{j \leftarrow i}} - \frac{r_j}{\sum\limits_{i \in I^j} c_{j \leftarrow i}} + \mu_1^j - \mu_2^j = 0, \forall i, j,$$
$$\mu_1^j(r_j - B_j^P) = 0, \forall j,$$
$$-\mu_2^j r_j = 0, \forall j, \tag{5.12}$$
$$\mu_1^j \geq 0, \forall j,$$
$$\mu_2^j \geq 0, \forall j.$$

Based on these KKT conditions, we have the following proposition about pricing:

Proposition 5.3 *Under competitive platforms, each platform j optimally announces its rewards to recruit participants by*

$$r_j^* = min\{B_j^P, \frac{\lambda_j}{2}\} \tag{5.13}$$

Proof Since the platform independently strategizes under a competitive platform scenario, we can consider specific platform j and extend the analysis to others. For the pricing of platform j, its optimal strategy can be solved from KKT conditions (5.12). Therefore, this proposition holds.   □

5.4 Detailed Design with Equilibrium Analysis: Cooperation Among Platforms

In this section, we study the interactions among inter-platforms that involve the concerns described in **CoopPricing**. For simplicity, we only consider the cooperation among specific platform j and its bargainers $\mathcal{B}^j$, since we can generate the analysis to any platform in $\mathcal{K}$. We will use Nash Bargaining Theory to explore the cooperations among j and $\mathcal{B}^j$, i.e., one-to-many bargaining framework [27]. Since the cooperation among all platforms needs to be determined, we apply the one-to-many bargaining framework to each platform in $\mathcal{K}$, which results in a many-to-many framework. Note that sharing sensing data does not mean exchanging data between both cooperative platforms.

When the bargainers of platform j are multiple (i.e., $N^j > 1$), it is important to guide bargaining proceeds (e.g., sequentially or concurrently) [29]. In this chapter, we consider sequential bargaining where platform j negotiates with $k \in \mathcal{B}^j$ one by one. The sequential bargaining has suitable and feasible solutions to characterize the studied problem from the algorithmic perspective. And the analysis of sequential bargaining can be extended into concurrent bargaining because it is proved that the payoffs of bargainers under concurrent bargaining are equal to the worst case of sequential bargaining [29]. Therefore, we pay attention to addressing the algorithmic challenges of bargaining and only consider sequential bargaining in this chapter.

5.4.1 Exact Bargaining

Supposing that platform j bargains with $\forall k \in \mathcal{B}^j$ in a given order $L^j[1 : N^j] = (l_1^j, l_2^j, \ldots, l_{N^j}^j)$, we analyze its Nash Bargaining Solutions (NBS), which is the cooperative outcome between platform j and $\forall k \in \mathcal{B}^j$, by backward induction from N^j to 1. That is, $(b_h^j, p_{h \leftarrow j})$ $(\forall h \in L^j[1 : s])$ is given in prior and $(b_g^{*j}, p_{g \leftarrow j}^*)$ $(\forall g \in L^j[s + 1 : N^j])$ has been induced, we predict $\left(b_{l_s^j}^j (b_{l[1:s-1]}^j), P_{l_s^j \leftarrow j}(b_{l[1:s-1]}^j) \right)$ by leveraging NBS, where $b_{l[1:s]}^j = (b_{l_1^j}^j, b_{l_2^j}^j, \ldots, b_{l_s^j}^j)$. Particularly, we also denote $p_{l[1:s] \leftarrow j} = (p_{1 \leftarrow j}, p_{2 \leftarrow j}, \ldots, p_{s \leftarrow j})$.

Now focusing on the subtlety of $\left(b_{l_s^j}^j (b_{l[1:s-1]}^j), P_{l_s^j \leftarrow j}(b_{l[1:s-1]}^j) \right)$, where platform j bargains with $s \in \mathcal{B}^j$ under given $b_{l[1:s-1]}^j$ and $b_{l[s+1:N^j]}^{*j}(b_{l_s^j}^j)$. Here, each $b_{l_k^j}^{*j} \in b_{l[s+1:N^j]}^{*j}(b_{l_s^j}^j)$ $(k > s)$ depends on the value determined in the first $k - 1$ stages, i.e., $b_{l_k^j}^{*j} | \left(b_{l[1:s-1]}^j, b_{l_s^j}^j, b_{l[s+1:N^j]-l_k^j}^{*j}(b_{l_s^j}^j) \right)$, where $b_{l[s+1:N^j]-l_k^j}^{*j}(b_{l_s^j}^j)$ is the induced cooperative vector excluding $b_{l_k^j}^{*j}$.

When platform j agrees with $l_s^j \in \mathcal{B}^j$ under the above situation, their payoffs are:

$$\Psi_j^{P+} \left(\left(b_{l[1:s-1]}^j, 1, b_{l[s+1:N^j]}^{*j}(1) \right), \right.$$

$$\left. \left(p_{l[1:s-1] \leftarrow j}, p_{l_s^j \leftarrow j}^*(b_{l[1:s-1]}^j), p_{l[s+1:N^j] \leftarrow j}^*(1) \right) \right) =$$

$$f \left(X_{j \leftarrow l_s^j} + \sum_{b_k^j \in (b_{l[1:s-1]}^j, b_{l[s+1:N^j]}^{*j}(1))} b_k^j X_{j \leftarrow k} \right) \tag{5.14}$$

$$- p_{l_s^j \leftarrow j}^*(b_{l[1:s-1]}^j) - \sum_{k \in (p_{l[1:s-1] \leftarrow j}, p_{l[s+1:N^j] \leftarrow j}^*(1))} k.$$

$$V_{l_s^j}^j \left(1, p_{l_s^j \leftarrow j}^*(b_{l[1:s-1]}^j) \right) = p_{l_s^j \leftarrow j}^*(b_{l[1:s-1]}^j) - E_{l_s^j} X_{j \leftarrow l_s^j}. \tag{5.15}$$

When platform j disagrees with $l_s^j \in \mathcal{B}^j$ under above situation, their payoffs are:

$$
\Psi_j^{P+}\left(\left(b_{l[1:s-1]}^j, 0, b_{l[s+1:N^j]}^{*j}(0)\right),\right.
$$

$$
\left.\left(p_{l[1:s-1]\leftarrow j}, 0, p_{l[s+1:N^j]\leftarrow j}^*(0)\right)\right) =
$$

$$
f\left(\sum_{b_k^j \in (b_{l[1:s-1]}^j, b_{l[s+1:N^j]}^{*j}(0))} b_k^j X_{j \leftarrow k}\right) \tag{5.16}
$$

$$
- \sum_{k \in (p_{l[1:s-1]\leftarrow j}, p_{l[s+1:N^j]\leftarrow j}^*(0))} k.
$$

$$
V_{l_s^j}^j(0, 0) = 0. \tag{5.17}
$$

For simplicity and ease of exposition, we denote l.h.s of (5.14) and (5.16) as
$\Psi_j^{P+}\left(1, p_{l_s^j \leftarrow j}(b_{l[1:s-1]}^j)\right)\big|_{(b_{l[1:s-1]}^j, b_{l[s+1:N^j]}^{*j})}^{p_{l[1:s-1]\leftarrow j}, p_{l[s+1:N^j]\leftarrow j}^*}$ and $\Psi_j^{P+}(0, 0)\big|_{(b_{l[1:s-1]}^j, b_{l[s+1:N^j]}^{*j})}^{p_{l[1:s-1]\leftarrow j}, p_{l[s+1:N^j]\leftarrow j}^*}$.

According to Nash Bargaining Theory [44], the NBS with respect to
$\left(b_{l_s^j}^{*j}(b_{l[1:s-1]}^j), p_{l_s^j \leftarrow j}^*(b_{l[1:s-1]}^j)\right)$, which is the product of two terms (Eq.(5.14)
minus Eq.(5.16) and Eq.(5.15) minus Eq.(5.17)), is derived from following
optimization problem **(CoopOPT)**:

$$
\max \quad \left(\Psi_j^{P+}\left(1, p_{l_s^j \leftarrow j}(b_{l[1:s-1]}^j)\right)\Big|_{(b_{l[1:s-1]}^j, b_{l[s+1:N^j]}^{*j})}^{p_{l[1:s-1]\leftarrow j}, p_{l[s+1:N^j]\leftarrow j}^*}\right.
$$

$$
\left. - \Psi_j^{P+}(0, 0)\Big|_{(b_{l[1:s-1]}^j, b_{l[s+1:N^j]}^{*j})}^{p_{l[1:s-1]\leftarrow j}, p_{l[s+1:N^j]\leftarrow j}^*}\right) \times \tag{5.18}
$$

$$
\left(V_{l_s^j}^j\left(1, p_{l_s^j \leftarrow j}(b_{l[1:s-1]}^j)\right) - V_{l_s^j}^j(0, 0)\right)
$$

$$
\text{s.t.} \quad \Psi_j^{P+}\left(1, p_{l_s^j \leftarrow j}(b_{l[1:s-1]}^j)\right)\Big|_{(b_{l[1:s-1]}^j, b_{l[s+1:N^j]}^{*j})}^{p_{l[1:s-1]\leftarrow j}, p_{l[s+1:N^j]\leftarrow j}^*}
$$

$$
- \Psi_j^{P+}(0, 0)\Big|_{(b_{l[1:s-1]}^j, b_{l[s+1:N^j]}^{*j})}^{p_{l[1:s-1]\leftarrow j}, p_{l[s+1:N^j]\leftarrow j}^*} \geq 0, \tag{5.19}
$$

$$V_{l_s^j}^j\left(1, P_{l_s^j \leftarrow j}(b_{I[1:s-1]}^j)\right) \geq 0, \tag{5.20}$$

$$P_{l_s^j \leftarrow j}(b_{I[1:s-1]}^j) \leq r_j^*. \tag{5.21}$$

In **CoopOPT**, $\left(\Psi_j^{P+}(0,0)\big|_{(b_{I[1:s-1]}^j, b_{I[s+1:Nj]}^{*j})}^{p_{I[1:s-1]\leftarrow j}, p_{I[s+1:Nj]\leftarrow j}^*}\right.$,

$V_{l_s^j}^j(0,0)\big)$ is disagreement point [27]. Constraints (5.19) and (5.20) indicate that NBS should deviate the bargaining outcome from the disagreement point where both payoffs should be greater than the disagreement point such that they have incentives to cooperate. Constraint (5.21) shows that the payment platform j gives to share sensing data from $l_s^j \in \mathcal{B}^j$ under cooperative platform scenario should be less rewards r_j^* priced in competitive platform scenario, otherwise it makes no sense to cooperate.

Leveraging KKT conditions, the NBS of **CoopOPT** can be obtained. Besides, we directly summarize the NBS in the following proposition due to limited space.

Proposition 5.4 *When platform j bargains with any $l_s^j \in \mathcal{B}^j$ sequentially under specific order $L^j[1 : N^j]$, j and l_s^j will reach an agreement on $\left(b_{l_s^j}^{*j}(b_{I[1:s-1]}^j), p_{l_s^j \leftarrow j}^*(b_{I[1:s-1]}^j)\right)$ if and only if:*

$$(b_{l_s^j}^{*j}, p_{l_s^j \leftarrow j}^*) = \begin{cases} (1, p_{l_s^j \leftarrow j}^*) & \begin{array}{l} if\ 0 \leq p_{l_s^j \leftarrow j}^* \leq min\{r_j^*, \Omega_s\} \\ and\ E_{l_s^j} \leq \dfrac{\Omega_s}{X_{j \leftarrow l_s^j}} \end{array} \\ (0,0) & otherwise, \end{cases} \tag{5.22}$$

where

$$p_{l_s^j \leftarrow j}^* = \frac{\alpha_s - \beta_s + \gamma_s + E_{l_s^j} X_{j \leftarrow l_s^j}}{2}, \tag{5.23}$$

$$\begin{cases} \alpha_s = f(X_{j \leftarrow l_s^j} + \displaystyle\sum_{b_k^j \in (b_{I[1:s-1]}^j, b_{I[s+1:Nj]}^{*j})^{(1)}} b_k^j X_{j \leftarrow k}), \\ \beta_s = f(\displaystyle\sum_{b_k^j \in (b_{I[1:s-1]}^j, b_{I[s+1:Nj]}^{*j})^{(0)}} b_k^j X_{j \leftarrow k}), \\ \gamma_s = \displaystyle\sum_{k \in p_{I[s+1:Nj]\leftarrow j}^{*(0)}} k - \displaystyle\sum_{k \in p_{I[s+1:Nj]\leftarrow j}^{*(1)}} k, \\ \Omega_s = \alpha_s - \beta_s + \gamma_s. \end{cases} \tag{5.24}$$

From Proposition 5.4, we can intuitively understand that if the price platform l_s^j charging to j is reasonable (non-negative but bound at the rewards used to recruit participants under a competitive scenario) and its sharing cost is not greater than a bound Ω, then they will agree to cooperate and share sensing data. We further summarize the insight about the bargaining between platform j and $\mathcal{B}^j$ in Algorithm 5.2:

Algorithm 5.2 SeqBargaining($s, b_{l[1:s-1]}^j$)

Input: recursive depth s, given bargaining outcome $b_{l[1:s-1]}^j$

Output: the NBS $(b_{l_s^j}^{*j}, p_{l_s^j \leftarrow j}^*)$ at stage s

1: **if** $s = N^j$ **then**
2:      Calculate $(b_{l_s^j}^{*j}, p_{l_s^j \leftarrow j}^*)$ using proposition 5.4
3:      return $(b_{l_s^j}^{*j}, p_{l_s^j \leftarrow j}^*)$
4: **end if**
5: $h_0 = 0, h_1 = 0, C = 0$
6: $pre_0 = (b_{l[1:s-1]}^j, 0)$
7: $pre_1 = (b_{l[1:s-1]}^j, 1)$
8: **for** $(i = s + 1; i \leq N^j; i + +)$ **do**
9:      $temp_0 = $ SeqBargaining(i, pre_0)
10:     $pre_0 = (pre_0, temp_0[0])$
11:     $h_0 = temp_0[0]X_{j \leftarrow l_i^j} + h_0$
12:     $temp_1 = $ SeqBargaining(i, pre_1)
13:     $pre_1 = (pre_1, temp_1[0])$
14:     $h_1 = temp_1[0]X_{j \leftarrow l_i^j} + h_1$
15:     $C = temp_0[1] - temp_1[1] + C$
16: **end for**
17: $A = f(X_{j \leftarrow l_s^j} + h_1)$
18: $B = f(h_1)$
19: $\Omega = A - B + C$
20: Calculate $p_{l_s^j \leftarrow j}^*$ using Eq. (5.23)
21: **if** $0 \leq p_{l_s^j \leftarrow j}^* \leq min\{r_j^*, \Omega\}$ and $E_{l_s^j} \leq \frac{\Omega}{X_{j \leftarrow l_s^j}}$ **then**
22:     return $(1, p_{l_s^j \leftarrow j}^*)$
23: **else**
24:     return $(0, 0)$
25: **end if**

When $s = N^j$, Algorithm 5.2 returns the NBS for stage s. Otherwise, it calculates two sub-problems in a recursive manner where one sub-problem corresponds to $b^{*j}_{l^j_s} = 0$ and the other corresponds to $b^{*j}_{l^j_s} = 1$. Finally, it aggregates these sub-problems to derive the NBS at stage $s < N^j$.

The essentials of Algorithm 5.2 is a backtracking algorithm. Its computing complexity is dominated by the following recursion formula:

$$
T(s) = \begin{cases} O(1) & s = N^j \\ \sum_{i=s+1}^{N^j} 2T(i) + O(1) & s < N^j. \end{cases}
\tag{5.25}
$$

By solving Eq. (5.25), we can derive $T(s) = (3^{N^j-1} - 1)(\frac{1}{3})^{s-1}$. In Algorithm 5.2, we have to calculate the result for each platform $s \in L^j[1:N^j]$ in each bargaining stage s. Therefore, the total complexity is the summation of $T(s)$ from $s = 1$ to N^j. That is, $\sum_{i=1}^{N^j} T(i) = \frac{(3^{N^j-1}-1)(1-(\frac{1}{3})^{N^j})}{1-\frac{1}{3}} = O(3^{N^j})$. Therefore, Algorithm 5.2 is computationally inefficient due to its exponential complexity.

The significantly exponential complexity motivates us to explore the NP-hardness of sequential bargaining we formulated in this chapter. In fact, the reason why Algorithm 5.2 is highly complex is that some sub-problem is repeatedly computed, i.e., existing overlapping sub-problem (the number of overlapping sub-problem is $\sum_{i=1}^{N^j-1} i = \frac{N^j(N^j-1)}{2}$). We cope with overlapping sub-problem in a memo manner. That is, using a memo to store the solution already solved. Before calculating the NBS in stage s, we can first check if the sub-problem in stage s has been solved and saved in the memo. If yes, we directly use the saved NBS in the memo; otherwise, we dive into the calculation of the current sub-problem. Although the number of sub-problem is polynomially bound and some complexity is reduced using the memo to some degree, the relatively large sub-problem is still significantly too complex to solve. Therefore, a heuristic algorithm to reduce the complexity of sequential bargaining is imperative and motivated to study in the next subsection.

5.4.2 Heuristic Bargaining

In order to reduce the complexity of sequential bargaining to a polynomial bound, we propose a heuristics bargaining which only considers the current bargaining and historical bargaining outcome. We define a set C^j which represents that platform

---

Algorithm 5.3 Heuristic Bargaining

Input: Bargaining set $\mathcal{B}^j, X_{j \leftarrow s}$ and E_s ($\forall s \in \mathcal{B}^j$)

Output: cooperative set C^j

1: **for** $l_s^j \in \mathcal{B}^j$ **do**

2:      calculating $\widetilde{A}, \widetilde{B}$ and $\widetilde{\Omega}$ based on Eq. (5.26)

3:      calculating $(\widetilde{b}_{l_s^j}^{*j}, \widetilde{p}_{l_s^j \leftarrow j}^{*})$ based on Proposition 5.4

4:      **if** $\widetilde{b}_{l_s^j}^{*j} = 1$ **then**

5:         $C^j = C^j \bigcup \{(l_s^j, \widetilde{p}_{l_s^j \leftarrow j}^{*})\}$

6:      **end if**

7: **end for**

8: **return** C^j

$k \in C^j$ cooperates with platform j, and then we refine the Eq. (5.24) as follows:

$$
\begin{cases}
\widetilde{\alpha}_s = f(X_{j \leftarrow l_s^j} + \displaystyle\sum_{l_k^j \in C^j \setminus l_s^j} X_{j \leftarrow l_k^j}), \\[4mm]
\widetilde{\beta}_s = f(\displaystyle\sum_{l_k^j \in C^j \setminus l_s^j} X_{j \leftarrow l_k^j}), \\[4mm]
\widetilde{\Omega}_s = \widetilde{\alpha}_s - \widetilde{\beta}_s.
\end{cases}
\tag{5.26}
$$

According to $\widetilde{\Omega}$ calculated in Eq. (5.26), we derive the bargaining outcome based on Proposition 5.4. If $\widetilde{b}_{l_s^j}^{*j} = 1$, then add l_s^j to C^j. We summarize the heuristic bargaining in Algorithm 5.3.

Algorithm 5.3 sequentially calculates $(\widetilde{b}_{l_s^j}^{*j}, \widetilde{p}_{l_s^j \leftarrow j}^{*})$ for each platform l_s^j and decides whether and how much to cooperate. The total process is done via the loop. Therefore, its complexity is $O(N^j)$. In the Heuristic Bargaining, we derive $(\widetilde{b}_{l_s^j}^{*j}, \widetilde{p}_{l_s^j \leftarrow j}^{*})$ using Proposition 5.4 as we do in Sequential Bargaining.

5.4.3   Many-To-Many Bargaining

In a cooperative platform scenario, cooperation represents the potential that one platform can share its sensing results collected from the competitive stage with another platform and benefit from this kind of cooperation. However, this sharing sensing is unilateral due to different data demands and data values. On the other hand, the bargaining framework proposed above is with respect to one-to-many rather than for any platforms. Therefore, many-to-many bargaining is necessary for a cooperative platform scenario.

Due to limited space, we only briefly illustrate the key idea of our many-to-many bargaining solution. First, many-yo-many bargaining determines the bargainers $\mathcal{B}^j$ for platform j. Second, it carries out one-to-many bargaining to determine their cooperation. Finally, this process repeats until the cooperation of all platforms is complete. When many-to-many bargaining integrates Algorithm 5.2, its complexity is $O(K3^{N^j})$ since K platforms run exact one-to-many bargaining. On the other hand, its complexity is polynomially dominated by $O(KN^j)$ when integrating Algorithm 5.3.

5.5 Performance Evaluation

5.5.1 Simulation Settings

We conduct simulations over two traces: random traces and D4D traces [45]. We generate two random traces, each for the competitive platform scenario and cooperative platform scenario, respectively. We set the trace delay $T_d = 24 \times 60 \times 60$. For each time slot t, we randomize the encounters between participants and platforms. We also randomize the encounters between each pair of platforms. We also conduct simulations over a real-world dataset: the Orange Data for Development (D4D) challenge dataset. We use SET1 and SET2 (sub-datasets of D4D) for the cooperative and competitive platform scenarios, respectively. SET1 consists of antenna-to-antenna encounters with call numbers and call duration where antennas stand for platforms as well as call numbers and duration are normalized to bargaining data amount and cost. SET2 consists of user call records, where the user stands for a participant and the antenna represents the platform. Due to no sensing cost and budget data along with SET2, we randomly generate them according to uniform distributions. The dataset time slot covers a total of 3600 hours from December 1, 2011 to April 28, 2012. There are more than 50,000 customers for 2 weeks and more than 1000 cellular towers in D4D. But we select the 30 most popular cellular towers as platforms and the 1000 most active customers as participants. We summarize the parameters in Table 5.2, where $U[\cdot]$ is a uniform distribution.

5.5.2 Simulation Results

In the competitive platform scenario, we explore how the user chooses the best platform to contribute and how to optimally exert sensing efforts $x^*_{j \leftarrow i}$. Here we evaluate how $x^*_{j \leftarrow i}$ is influenced by some factors. We sample r_j, B^U_i and $c_{j \leftarrow i}$ according to Table 5.2. Furthermore, $x^*_{j \leftarrow i}$ is plotted based on Proposition 5.1. Figure 5.2 shows that rewards r_j announced by platform j, efforts bound B^U_i and sensing cost $c_{j \leftarrow i}$ greatly affect the sensing efforts $x^*_{j \leftarrow i}$ user i exerting to platform

Table 5.2 Simulation parameters

Parameter		Value
Competitive scenario	r_j	$\sim U[0, 200]$
	$c_{j \leftarrow i}$	$\sim U[2, 9]$
	B_i^U	$\sim U[10, 60]$
	λ_j	$\sim U[0, 200]$
	B_j^P	$\sim U[0, 200]$
Cooperative scenario	$X_{j \leftarrow k}$	$\sim U[1, 20]$
	E_k	$\sim U[10, 30]$
	p	0.4
	r_j^*	200
Number of participants	I	1000
Number of platform	K	30

Fig. 5.2 The impact of rewards (r_j) offered by the platform and users' sensing budget (B_i^U) on the user's optimal efforts ($x_{j \leftarrow i}^*$)

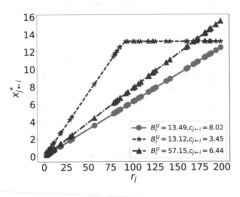

Fig. 5.3 The effect of rewards (r_j) offered by the platform and user's sensing costs ($c_{j \leftarrow i}^*$) on user's optimal efforts ($x_{j \leftarrow i}^*$) given users' certain sensing budget

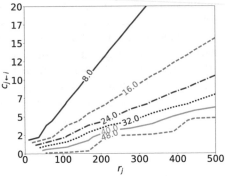

j. $x_{j \leftarrow i}^*$ monotonously increases with r_j but decreases with B_i^U and $c_{j \leftarrow i}$, which is also reflected in Fig. 5.3. Besides, we can also observe $c_{j \leftarrow i}$ is non-linear to $x_{j \leftarrow i}^*$.

From the perspective of platforms, they should subtly price on rewards in order to compete for more participants and sensing data. We have analyzed the pricing policy based on maximizing their payoffs in Proposition 5.3. Here we evaluate how r_j^* is influenced by some factors. We sample B_j^P and λ_j according to Table 5.2 as well. It

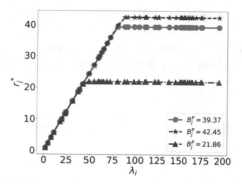

Fig. 5.4 The effect of data revenues (λ_j) and rewarding budget (B_j^P) of the platform on the priced rewards (r_j^*)

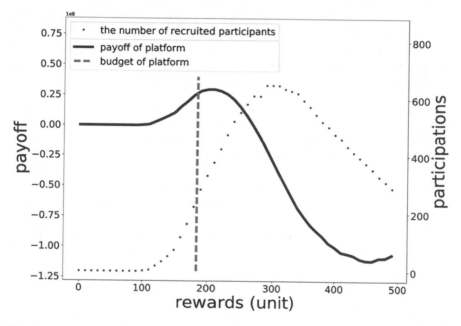

Fig. 5.5 The impact of the offered rewards (r_j) and rewarding budget (B_j^P) of the platform on a specific platform (payoff and participant number)

is easy to see in Fig. 5.4 that λ_j positively affects r_j^* within B_j^P. On the other hand, λ_j has the same impact on announced rewards r_j^* for all platforms. The announced rewards r_j^* will increase unless it reaches B_j^P which is the pricing budget.

In Fig. 5.5, we generally evaluate one specific platform's payoff and the number of participants it recruits over random trace (results can be extended to D4D because they have the same trend) in a competitive platform scenario. It is shown that the payoff first increases with r_j, but then decreases (concave shape) That is, certain sensing efforts from participants bring higher Service of Value (SoV) $\Phi_j(x^j)$, which positively affects the platform's payoff. But over-pricing on rewards r_j will result in higher costs for the platform. As we proved in Proposition 5.1, the platform's payoff is concave on r_j. Besides, the green line represents the platform's rewarding budget. This motivates the platform to set as much budget according to maximum point r_j^* as possible. The number of participants that which platform recruits have the same trend with the platform's payoff under r_j. The reason why the number of participants goes down in higher r_j is that higher r_j requires the user to exert more sensing efforts $x_{j\leftarrow i}^*$. However, each user has sensing efforts bound to B_i^U (such as the remaining battery level). This again verifies that the over-rewarding platform is not necessarily the best platform for all users. Interestingly, the payoff maximum does not coincide with the maximum of the number of participants. This again verifies the effect of multiple platforms because other platforms' optimal rewards can competitively attract participants.

In Fig. 5.6, we evaluate how the number of bargainers affects social welfare in a cooperative platform scenario and the difference between different bargaining methods in social welfare. Here, we compare Sequential bargaining (Algorithm 5.2) and Heuristic bargaining (Algorithm 5.3). In addition, we use Random bargaining which calculates payment $p_{l_s^j\leftarrow j}^*$ according to Eq. (5.23) and set $b_{l_s^j}^{*j} = 1$ by probability (using 0.4 in simulation) as a benchmark. Let N^j range in [1, 15]. For each loop, the bargaining outcome is derived by running Algorithm 5.2, Algorithm 5.3 and Random bargaining. Then social welfare is calculated according to the bargaining outcome in each loop, respectively. Clearly, under Sequential bargaining and Heuristic bargaining, the larger number of bargainers is, the higher social welfare is. This is because a large number of bargainers result in a higher probability and more opportunity to cooperate. Moreover, the performance of Heuristic bargaining approaches to Sequential bargaining by about 80% on average. Besides, Random bargaining suffers from the uncertainty of selecting cooperators,

Fig. 5.6 The relationship between the number of bargainers and cooperative payoff of the platform

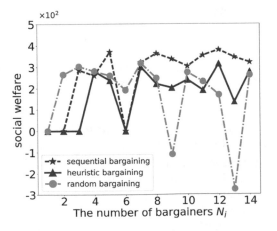

Fig. 5.7 The bargainer distribution under sequential bargaining where E is bargaining cost and X is the number of sensing data

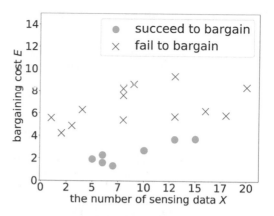

which results in an unstable payoff. Therefore, Sequential bargaining and Heuristic bargaining obviously outperform Random bargaining.

In Fig. 5.7, we investigate the cooperative distribution of bargainers for a specific platform that holds bargaining for requesting sensing data. Cooperative distribution is evaluated from two dimensions: sensing data for bargaining X and bargaining cost E. We set $N^j = 20$ and sample X and E according to Table 5.2. Sequential bargaining (Algorithm 5.2) is run to derive the NBS. From NBS, we can determine the number of cooperators and which one to cooperate with. And we find that cooperators are distributed into the area where X takes a larger value but E takes a smaller value.

5.6 Conclusion

In this chapter, we study the competitive and cooperative mechanisms for multiple platforms in mobile crowdsensing. The competitive platform scenario is formulated as a two-stage Stackelberg game. Nash equilibrium indicates how platforms and users optimally make decisions under a multiple platform environment. We further study the cooperative mechanism between platforms for the cooperative platform scenario. It is formulated as a many-yo-many bargaining framework. We solve the many-to-many bargaining framework by combining one-to-many bargaining. Numerical results show that our proposed competitive and cooperative mechanisms not only guarantee the platforms' payoff by optimally pricing on rewards and guiding participants to select the best platform, but also further improve platforms' payoff by addressing how platforms bargain to cooperatively share sensing data.

References

1. Guo, B., Wang, Z., Yu, Z., Wang, Y., Yen, N.Y., Huang, R., Zhou, X.: Mobile crowd sensing and computing: The review of an emerging human-powered sensing paradigm. ACM Comput. Surv. (CSUR) **48**(1), 7 (2015)
2. Jarrett, J., Blake, M.B., Saleh, I.: Crowdsourcing, mixed elastic systems and human-enhanced computing–a survey. IEEE Trans. Serv. Comput. **11**(1), 202–214 (2018)
3. Vasserman, S., Michal, F., Avinatan, H.: Implementing the wisdom of waze. In: Twenty-Fourth International Joint Conference on Artificial Intelligence (IJCAI), pp. 660–666, AAAI Press (2015)
4. Duan, S., Lai, J., Link, M.: Crowdsourcing via mobile evaluating viability of data collection "Gigs" with iPhone users. Surv. Pract. **6**(3) (2013)
5. Zheng, Y., Liu, F., Hsieh, H.-P.: U-air: When urban air quality inference meets big data. In: ACM SIGKDD, pp. 1436–1444 (2013)
6. Guo, B., Chen, H., Yu, Z., Xie, X., Huangfu, S., Zhang, D.: Fliermeet: a mobile crowdsensing system for cross-space public information reposting, tagging, and sharing. IEEE TMC **14**(10), 2020–2033 (2015)
7. Zhang, X., Yang, Z., Sun, W., Liu, Y., Tang, S., Xing, K., Mao, X.: Incentives for mobile crowd sensing: A survey. IEEE Commun. Surv. Tutorials **18**(1), 54–67 (2016)
8. Yang, D., Xue, G., Fang, X., Tang, J.: Crowdsourcing to smartphones: incentive mechanism design for mobile phone sensing. In: ACM MobiCom (2012)
9. Hu, C., Xiao, M., Huang, L., Gao, G.: Truthful incentive mechanism for vehicle-based nondeterministic crowdsensing. In: IEEE IWQoS (2016)
10. Gao, L., Hou, F., Huang, J.: Providing long-term participation incentive in participatory sensing. In: IEEE INFOCOM (2015)
11. Chen, Y., Li, B., Zhang, Q.: Incentivizing crowdsourcing systems with network effects. In: IEEE INFOCOM (2016)
12. Jin, H., Su, L., Nahrstedt, K.: Centurion: Incentivizing multi-requester mobile crowd sensing. In: IEEE INFOCOM (2017)
13. Xie, N., Lui, J.C.S.: Incentive mechanism and rating system design for crowdsourcing systems: analysis, tradeoffs and inference. IEEE Trans. Serv. Comput. **11**, 90–102 (2016)
14. Zhan, Y., Xia, Y., Li, Y., Li, F., Wang, Y.: Time-sensitive data collection with incentive-aware for mobile opportunistic crowdsensing. IEEE TVT **66**(6), 5301–5315 (2017)
15. Wu, Y., Li, F., Ma, L., Xie, Y., Li, T., Wang, Y.: A context-aware multi-armed bandit incentive mechanism for mobile crowd sensing systems. IEEE IoT-J **6**, 7648–7658 (2019)
16. Li, T., Jung, T., Qiu, Z., Li, H., Cao, L., Wang, Y.: Scalable privacy-preserving participant selection for mobile crowdsensing systems: Participant grouping and secure group bidding. IEEE Trans. Network Sci. Eng. **7**, 855–868 (2019)
17. Scekic, O., Truong, H.-L., Dustdar, S.: Modeling rewards and incentive mechanisms for social bpm. In: International Conference on Business Process Management, pp. 150–155. Springer (2012)
18. Wang, Y., Cai, Z., Yin, G., Gao, Y., Tong, X., Wu, G.: An incentive mechanism with privacy protection in mobile crowdsourcing systems. Comput. Netw. **102**, 157–171 (2016)
19. Duan, Z., Li, W., Cai, Z.: Distributed auctions for task assignment and scheduling in mobile crowdsensing systems. In: IEEE ICDCS, pp. 635–644 (2017)
20. Yu, H., Cheung, M.H., Gao, L., Huang, J.: Economics of public Wi-Fi monetization and advertising. In: IEEE INFOCOM (2016)
21. Wen, Y., Shi, J., Zhang, Q., Tian, X., Huang, Z., Yu, H., Cheng, Y., Shen, X.: Quality-driven auction-based incentive mechanism for mobile crowd sensing. TVT **64**(9), 4203–4214 (2015)
22. Zhang, X., Xue, G., Yu, R., Yang, D., Tang, J.: Truthful incentive mechanisms for crowdsourcing. In: IEEE INFOCOM (2015)
23. Wang, Y., Cai, Z., Tong, X., Gao, Y., Yin, G.: Truthful incentive mechanism with location privacy-preserving for mobile crowdsourcing systems. Comput. Netw. **135**, 32–43 (2018)

24. Liu, Y., Guo, B., Wang, Y., Wu, W., Yu, Z., Zhang, D.: Taskme: multi-task allocation in mobile crowd sensing. In: ACM UbiComp, pp. 403–414 (2016)
25. Li, H., Li, T., Wang, W., Wang, Y.: Dynamic participant selection for large-scale mobile crowd sensing. In: IEEE TMC (2019)
26. Jarrett, J., Blake, M.B.: Towards a distributed worker-job matching architecture for crowdsourcing. In: IEEE WETICE, pp. 9–11 (2016)
27. Yu, H., Cheung, M.H., Huang, J.: Cooperative Wi-Fi deployment: a one-to-many bargaining framework. IEEE TMC **16**(6), 1559–1572 (2017)
28. Zhang, M., Gao, L., Huang, J., Honig, M.: Cooperative and competitive operator pricing for mobile crowdsourced internet access. In: IEEE INFOCOM (2017)
29. Gao, L., Iosifidis, G., Huang, J., Tassiulas, L., Li, D.: Bargaining-based mobile data offloading. IEEE J. Sel. Areas Commun. **32**(6), 1114–1125 (2014)
30. Yu, H., Iosifidis, G., Huang, J., Tassiulas, L.: Auction-based coopetition between LTE unlicensed and Wi-Fi. IEEE J. Sel. Areas Commun. **35**(1), 79–90 (2017)
31. Li, J., Cai, Z., Wang, J., Han, M., Li, Y.: Truthful incentive mechanisms for geographical position conflicting mobile crowdsensing systems. IEEE Trans. Comput. Social Syst. **5**(2), 324–334 (2018)
32. Dustdar, S., Guo, Y., Satzger, B., Truong, H.-L.: Principles of elastic processes. IEEE Internet Comput. **15**(5), 66–71 (2011)
33. Dustdar, S., Truong, H.-L.: Virtualizing software and humans for elastic processes in multiple clouds–a service management perspective. Int. J. Next-Gener. Comput. **3**(2) (2012)
34. Hoenisch, P., Schulte, S., Dustdar, S.: Workflow scheduling and resource allocation for cloud-based execution of elastic processes. In: 2013 IEEE 6th International Conference on Service-Oriented Computing and Applications, pp. 1–8 (2013)
35. Yu, J., Cheung, M.H., Huang, J., Poor, H.V.: Mobile data trading: Behavioral economics analysis and algorithm design. IEEE J. Sel. Areas Commun. **35**(4), 994–1005 (2017)
36. Ma, Q., Liu, Y.-F., Huang, J.: Time and location aware mobile data pricing. IEEE TMC **15**(10), 2599–2613 (2016)
37. Goodchild, M.F., Glennon, J.A.: Crowdsourcing geographic information for disaster response: a research frontier. Int. J. Digital Earth **3**(3), 231–241 (2010)
38. Scekic, O, Truong, H.-L., Dustdar, S: Incentives and rewarding in social computing. Commun. ACM **56**(6), 72–82 (2013)
39. Scekic, O., Truong, H.-L., Dustdar, S.: Programming incentives in information systems. In: International Conference on Advanced Information Systems Engineering, pp. 688–703. Springer (2013)
40. Jarrett, J., Blake, M.B.: Collaborative infrastructure for on-demand crowdsourced tasks. In: IEEE WETICE, pp. 9–14 (2015)
41. Cheung, M.H., Hou, F., Huang, J.: Make a difference: diversity-driven social mobile crowdsensing. In: IEEE INFOCOM (2017)
42. Johnson, D.S., Michael, R.G.: Computers and Intractability: A Guide to the Theory of NP-Completeness. WH Freeman, New York (1979)
43. Chong, E.K.P., Zak, S.H.: An Introduction to Optimization, vol. 76. John Wiley & Sons, New York (2013)
44. Nash, J.F. Jr.: The bargaining problem. Econometrica: J. Econ. Soc. **18**, 155–162 (1950)
45. Blondel, V.D., Esch, M., Chan, C., Clérot, F., Deville, P., Huens, E., Morlot, F., Smoreda, Z, Ziemlicki, C.: Data for Development: The D4D Challenge on Mobile Phone Data. CiteSeer, University Park (2012). Preprint, arXiv:1210.0137

Chapter 6
Summary

Abstract In this chapter, we summarize this book and discuss the future directions for incentive mechanisms in MCS.

Keywords Contributions · Open directions

6.1 Summary of the Book

As a distributed problem-solving paradigm, crowdsourcing has succeeded in many fields. When introducing crowdsourcing into the sensing tasks of the Internet of Things, combined with the characteristics of mobile smart devices, it gives rise to a human-centric sensing and computing paradigm, called mobile crowdsensing (MCS). MCS takes mobile devices as the basic perception unit, distributes sensing tasks to users through the platform, and allows users to collect multi-modal sensing data around them via their built-in sensors. MCS makes up for the shortcomings of WSNs, eliminating the need for large-scale deployment of professional sensors and greatly saving deployment costs. The mobility and autonomy of users make MCS easy to maintain, with strong interoperability, scalability, and flexibility.

Although MCS has revolutionized traditional WSNs, its own characteristics have created many new problems. One of the key issues is that rational users have selfish tendencies and are unwilling to participate in sensing tasks or contribute sensing data if they don't receive appropriate rewards. On the one hand, executing sensing tasks inevitably consumes the resources of the user's device, such as battery power, storage, and computing power, and transmitting sensing data incurs additional communication costs. On the other hand, sensing tasks are often location-related, which raises security concerns about exposing the user's location privacy. These issues have increased the barriers to user participation in MCS and lowered their enthusiasm for participation. Therefore, designing appropriate incentive mechanisms to ensure that users contribute sensing data to MCS while receiving real benefits is one of the core issues in MCS research. By using reasonable rewards to obtain high-quality sensing data and achieve a virtuous cycle of "all for one, one for all" in the MCS system, we can address this challenge.

© The Author(s), under exclusive license to Springer Nature Singapore Pte Ltd. 2024   125
Y. Li et al., *Incentive Mechanism for Mobile Crowdsensing*, SpringerBriefs
in Computer Science, https://doi.org/10.1007/978-981-99-6921-0_6

Incentive mechanisms are typically designed using game theory, which involves designing game rules that make it advantageous for users to participate in sensing tasks while maximizing the overall benefits of the sensing system subject to specific constraints on the requester and user. However, different scenarios present different challenges, and the design goals of incentive mechanisms need to be readjusted and new factors need to be considered, such as limited budgets, insufficient user participation, unknown user sensing quality information, fairness requirements for user selection, platform profit goals, and coexistence of competition and cooperation among multiple platforms. In order to address these challenges, this book focuses on designing effective incentive mechanisms for MCS in different scenarios by leveraging game theory, and makes the following contributions:

1. **A long-term incentive mechanism with profit and sustainability guarantee**. To prevent user attrition and ensure the platform profits from coordinating between requesters and users, we have designed an incentive mechanism to achieve these two goals. We use a three-stage Stackelberg game to model the incentive mechanism, and aim at addressing issues: (1) how to ensure that users are fairly selected as workers, thus ensuring their long-term participation in MCS; (2) how to calculate users' interest in task subsets based on their computing capabilities; (3) and how to ensure that the sensing platform is profitable on average over time, thereby ensuring the sustainable development of the MCS system. In each round of strategy interaction, we analyzed the equilibrium strategies of the three-stage Stackelberg game. In stage I, we combined Zinkevich's online gradient learning method with the drift-plus-penalty technique of Lyapunov optimization to design an online task pricing algorithm. This algorithm achieves a tradeoff performance between platform revenue and sensing system efficiency (more users executing tasks) within the range $[O(1/v), O(v)]$. In stage II, we designed an FPTAS algorithm to compute, for each user with different computing capabilities, a set of $(1 - \varepsilon)$-optimal tasks as their interest information to submit to the platform. In stage III, the platform designs an online user selection algorithm based on the interest information reported by users, by combining an approximation algorithm of set cover and the Lyapunov optimization technique. This algorithm achieves a tradeoff performance between ensuring user long-term selection and minimizing the platform's recruitment rewards within the range of $[O(1/V), O(V)]$. Finally, we conduct simulation experiments to validate the performance of the incentive mechanism in terms of pricing rationality, user selection fairness, and effectiveness.

2. **A fair incentive mechanism with privacy preservation and quality guarantee**. To address the issues of unknown user quality information and fairness requirements in user selection, we studied the incentive mechanism design problem for ensuring quality perception fairness in dynamic MCS scenarios. We modeled the strategic interaction between the platform and users in the incentive mechanism as a three-stage Stackelberg game. In each round of strategy interaction, the platform, in the first stage, devises the optimal rewards satisfying the incentive budget for each arriving task to achieve the user participation rate

requirement. In the second stage, each user reports a set of tasks as their interest set to the platform based on the released task set and corresponding rewards. In the third stage, the platform selects high-quality users to complete all the tasks after receiving the interest set information from all users. As user quality information is unknown, we use a combinatorial sleeping multi-armed bandit model to design a user selection algorithm that learns the estimated perception quality information based on multi-round decisions. To ensure fairness, we create a virtual credit queue for all users and design a user selection algorithm using the Lyapunov optimization technique to achieve the balance of virtual credit income and expenditure for ensuring fairness in user selection. We prove that the user selection algorithm based on the combinatorial sleeping multi-armed bandit model achieves a sublinear time regret of $O(\sqrt{b|N|T\log T})$ in learning user quality information, and the Lyapunov optimization user selection algorithm oriented toward virtual credits achieves a performance tradeoff range of $[O(1/V), O(V)]$ in the two conflicting objectives of optimal user selection and fairness guarantee. Finally, we evaluate the effectiveness of the proposed incentive mechanism in terms of pricing rationality, asymptotic optimality in learning user quality information, and fairness in user selection through simulation experiments.

3. **A collaborative incentive mechanism with POI-tagging App assistance.** To address the problem of insufficient user participation, we consider the assistance of third-party applications in the incentive mechanism design of crowdsensing and use a three-stage Stackelberg game to model the incentive mechanism design problem under this third-party application assistance. We analyze the equilibrium strategies of the three-stage Stackelberg game and, in the first stage, the third-party application determines the interest point tagging price to maximize its own benefits. In the second stage, the sensing platform determines how to choose a set of tasks as interest points for the third-party application to tag based on the known tagging price, thereby increasing the number of task participants. In the third stage, ordinary users and app users decide whether to perform sensing tasks and which optimal task to choose to execute. Finally, we evaluate the role of third-party application assistance in the incentive mechanism and reveal the rationality of pricing rewards and the efficiency of recruiting users with the help of experiments and simulations.

4. **A coopetition-aware incentive mechanism with multiple platform coexistence.** For the multi-platform MCS scenario, we identify the challenges of designing incentive mechanisms for multi-platform coexistence and divide multi-platform coexistence into competitive and cooperative relationships. For the competitive relationship between platforms, we model the incentive mechanism as a Stackelberg game with multiple leaders and multiple followers. By deriving the Stackelberg equilibrium, each platform can price tasks based on the computed optimal rewards, and each user can choose the best platform to perform sensing tasks. For the cooperative relationship between platforms, we use the Nash bargaining model in cooperative games to model the incentive mechanism for multi-platform cooperation scenarios. By first proving the NP-

hardness of the one-to-many Nash bargaining model, we propose a heuristic bargaining solution. Deriving the Nash bargaining solution reveals how platforms can cooperate to share sensing data and how much reward to offer for cooperation, enabling secondary use of sensing data. Finally, we evaluate the performance of the proposed incentive mechanism for incentivizing multi-platform coexistence through experiments and simulations, demonstrating that the incentive mechanism can not only solve the strategic interaction between platforms and participants but also between platforms.

6.2 Future Directions

Game-theoretic incentive mechanisms have been well-designed and developed to address many challenges in crowdsensing. When game-theoretic incentive mechanisms focus on how to price appropriate rewards based on the game-theoretic approaches, however, it remains some open issues that need to be further explored and studied in the future. In this section, we pay attention to discussing and summarizing the future directions in the game-theoretic incentive mechanism design.

1. **Integration of Multiple Game Models in Incentive Mechanisms.** Existing research on incentive mechanisms for MCS mostly adopts a certain game model to model the strategic interaction between requesters, platforms, and users. These game models include Stackelberg games, auction theory, Nash bargaining, and contract theory, among others. Different game models have their own characteristics and address different challenges, resulting in different fundamental principles for the designed incentive mechanisms and different operating modes for MCS systems. To comprehensively consider the influence of various challenging factors in the actual operating environment of crowdsensing and design incentive mechanisms that integrate multiple game models to address various problems in MCS, it is important to develop a universal MCS incentive mechanism that can adapt to different scenarios.

2. **Methods for Solving Equilibrium Points in Model-Free Games.** Many existing research works on incentive mechanisms based on game models often assume convex utility functions for platforms and users, and derive unique equilibrium point interaction strategies. However, in practical situations, the equilibrium points derived by assuming utility function models may not reflect the optimal strategies of the incentive mechanisms. Different fields of MCS applications may have utility functions in different forms, and the constant parameters in the game model may not be known. Therefore, studying more general model-free games is an important challenge in designing incentive mechanisms for crowdsensing. As machine learning algorithms such as supervised learning, reinforcement learning, and unsupervised learning provide some inspiration for solving equilibrium points in model-free games, it has become a meaningful research direction for incentive mechanism research.

3. **Incentive Mechanisms for a Blockchain-Enabled Crowdsensing Architecture.** In practical situations, the issue of privacy and security is crucial to the feasibility of incentive mechanisms. Firstly, the main idea of incentive mechanisms is to compensate users for their privacy costs through rewards, which does not truly protect users' privacy data. This limitation restricts the application of crowdsensing in scenarios with high-security requirements. Secondly, the architecture of an MCS system often relies on a platform deployed on a cloud server, which serves as a central organization to coordinate the supply and demand relationship between requesters and users for data. However, this centralized platform has scalability issues with centralized task allocation and reward pricing, and can become overloaded with a large number of requesters and tasks. Finally, incentive mechanisms require assumptions of trust between the platform and users, and between the platform and requesters. The fees announced by requesters and the rewards determined by the platform must be honored after receiving sensing data. However, this mutual trust assumption may not always hold true in practical situations. With the emergence and development of blockchain technology, these challenges can be addressed. Blockchain technology is a new type of distributed "ledger" database that combines peer-to-peer communication protocols, consensus mechanisms, asymmetric encryption, blockchain structures, and smart contracts to achieve process trustworthiness and decentralization. Process trustworthiness means that operations are tamper-proof, traceable, and maintained by multiple parties, enabling multiple parties to establish trust at a low cost. Decentralization means that there is no single coordinator to maintain the synchronization of data and actions among multiple parties in the network, and network nodes rely entirely on peer-to-peer communication technology and consensus mechanisms to achieve synchronization of data and actions. In blockchain, data is transmitted through asymmetric encryption to achieve privacy protection, and smart contracts complete the operation of data in the block in a trustworthy manner through event triggers. The operation of the blockchain is then synchronized among network nodes through the consensus mechanism. However, compared with traditional crowdsensing, the system architecture of blockchain-based crowdsensing has significant differences. The decentralized nature means that the platform is optional, and task publishing, data collection, and reward transactions are all performed on the blockchain. Smart contracts become the main executor of the data supply and demand relationship between requesters and users. This new interaction method renders the existing incentive mechanisms inapplicable. Therefore, designing incentive mechanisms that are suitable for the new architecture and accelerating the collection of sensing data is one of the important research issues in blockchain-based crowdsensing.